T0213272

SpringerBriefs in Environmental Science

More information about this series at http://www.springer.com/series/8868

Daniel Żarski • Ákos Horváth
Gergely Bernáth • Sławomir Krejszeff
János Radóczi • Katarzyna Palińska-Żarska
Zoltán Bokor • Krzysztof Kupren
Béla Urbányi

Controlled Reproduction of Wild Eurasian Perch

A hatchery manual

Daniel Żarski
Department of Lake and River Fisheries
University of Warmia and Mazury
Olsztyn, Poland

Gergely Bernáth
Department of Aquaculture
Szent István University
Gödöllő, Hungary

János Radóczi
Szabolcsi Halászati Kft.
Nyíregyháza, Hungary

Zoltán Bokor
Department of Aquaculture
Szent István University
Gödöllő, Hungary

Béla Urbányi
Department of Aquaculture
Szent István University
Gödöllő, Hungary

Ákos Horváth
Department of Aquaculture
Szent István University
Gödöllő, Hungary

Sławomir Krejszeff
Department of Lake and River Fisheries
University of Warmia and Mazury
Olsztyn, Poland

Katarzyna Palińska-Żarska
Department of Lake and River Fisheries
University of Warmia and Mazury
Olsztyn, Poland

Krzysztof Kupren
Department of Tourism, Recreation and
   Ecology
University of Warmia and Mazury in
   Olsztyn
Olsztyn, Poland

ISSN 2191-5547   ISSN 2191-5555   (electronic)
SpringerBriefs in Environmental Science
ISBN 978-3-319-49375-6   ISBN 978-3-319-49376-3   (eBook)
DOI 10.1007/978-3-319-49376-3

Library of Congress Control Number: 2016961293

Printed on acid-free paper

This Springer imprint is published by Springer Nature
The registered company is Springer International Publishing AG
The registered company address is: Gewerbestrasse 11, 6330 Cham, Switzerland

Acknowledgements

The authors would sincerely like to thank Prof. Patrick Kestemont (University of Namur, Belgium) and Prof. Pascal Fontaine (University of Lorraine, France) for their careful review and suggestions which had considerably contributed to the final content and shape of the book.

Funding Information

The preparation of this book was funded by the project, Development of an induced spawning technology and hatchery manual for the Eurasian perch (*Perca fluviatilis*) (E!8028 PERCAHATCH), and financed under the EUREKA funding programme.

The project was implemented by Szabolcsi Halászati Kft. (Ltd.) (Nyíregyháza, Hungary) in collaboration with Szent István University (Gödöllő, Hungary) (funded by the National Research, Development and Innovation Fund of Hungary within the project No. EUREKA_HU_12-1-2012-0056) and the University of Warmia and Mazury (Olsztyn, Poland). Further financial support to the authors was provided by the COST Action 1205 AQUAGAMETE and Research Centre of Excellence – 11476-3/2016/FEKUT.

Contents

The European freshwater aquaculture production level has been stagnating for the last two decades, whereas marine and brackish-based production exhibits a constantly growing trend (Fig. 1.1). This stems from the fact that over 75 % of freshwater aquaculture production relies only on two species: common carp, *Cyprinus carpio* (L.), and rainbow trout, *Oncorhynchus mykiss* (Walbaum), whose production level has been stagnating since 1995 and oscillating around 350 thousand tonnes per year (FAO statistics). The main reason for this is probably the relatively constant market demand for the two species, which must compete with growing supply of high-quality products originating from marine-based aquaculture. In order to face the rapidly growing competition of mariculture, diversification of inland production is urgently needed (Fontaine 2004; Fontaine et al. 2009). However, traditional production systems (earthen-ponds and flow-through 'trout' farms) were found to be insufficient to effectively produce other species than those typically grown in them. In this regard, recirculating aquaculture systems (RAS) were found to be an excellent tool allowing intensive farming of species whose production was highly limited or even impossible in traditional inland fish farms. These systems allow a sustainable increase of the production level (through reduced water consumption, efficient waste management and with high hygiene and disease management) close to the target market (Martins et al. 2010). Unfortunately, although the RAS technology is known and has been developed for decades development in this area of aquacultural production is still very slow. Most recently, it was suggested that the structure of the freshwater aquaculture industry (consisting mostly of small companies) limits the innovation and investment in new technologies, being an indispensable element of development of this type of production, which requires high investments (Nielsen et al. 2015) and constant development of the production technologies (for more details on current status and prospects of Eurasian percid aquaculture we highly recommend to see also Overton et al. 2015 and Steenfeldt et al. 2015). On the other hand, the production in RAS is also limited due to the lack of clear and standardized production protocols allowing to fully utilize the high potential of this type of production technology. This concerns also controlled reproduction protocols being the

© The Authors 2017
D. Żarski et al., *Controlled Reproduction of Wild Eurasian Perch*, SpringerBriefs
in Environmental Science, DOI 10.1007/978-3-319-49376-3_1

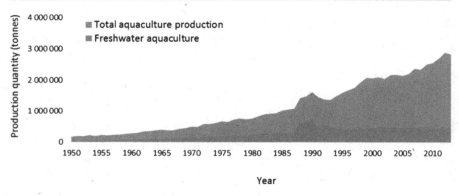

Fig. 1.1 European aquaculture production (total and freshwater) according to FAO statistics

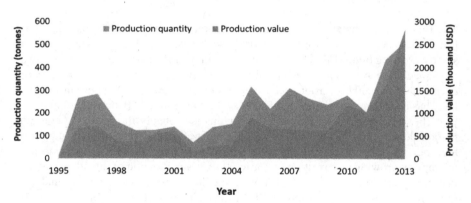

Fig. 1.2 Production quantity and value of Eurasian perch (According to FAO statistics)

first step toward the effective production and the basis for any further step of the culture process.

Among the freshwater species dedicated for the diversification of European freshwater aquaculture Eurasian perch, *Perca fluviatilis* L., was pointed out as one of the most promising emerging candidate (Fontaine et al. 2009; Steenfeldt et al. 2015). Its global production amounts to over 29 thousand tonnes in 2013 (FAO statistics). While wild catches remained stable for about 30 years, aquaculture production has exhibited a growing trend and reached, for the first time, over 500 tonnes in 2013 (FAO statistics) with a production value of over 2.8 million USD and determined average price of about 5.5 USD per kg (Fig. 1.2, FAO statistics). This clearly demonstrates that intensive aquaculture of this species is in the beginning of its development path and that the market demand is still insatiate (Fontaine et al. 2009; Overton et al. 2015). The European market for this species was very extensively described also by Toner (2015).

Biological (e.g. low growth rate under natural thermal conditions, sensitivity to water quality and pathogens) and behavioral features of Eurasian perch were found to be limiting factors during the intensive production in traditional (open and semi-open) production systems. However, it was found that these factors do not pose much of a problem while reared in RAS (Fontaine 2004). This led to intensive studies on controlled reproduction (Kucharczyk et al. 1996b; Kouril et al. 1997; Sulistyo et al. 1998; Kucharczyk et al. 1998) and aspects of intensive production in RAS (Fontaine et al. 1996, 1997). The first successful attempts determined the need for the development of protocols of intensive larval rearing (Mélard et al. 1996) and possibilities of feeding with compound diets (Kestemont et al. 1996; Fiogbé et al. 1996). At the beginning of XXI century various aspects of intensive larviculture (Abi-Ayad et al. 2000; Tamazouzt et al. 2000; Cuvier-Péres et al. 2001; Baras et al. 2003) as well as juvenile production technology (Xu and Kestemont 2002; Mandiki et al. 2004) were investigated. Also, the first attempts had been made to grow the spawners of Eurasian perch under controlled conditions (Jourdan et al. 2000) and manipulate environmental conditions enabling control of the reproductive cycle (Migaud et al. 2002; Wang et al. 2006). Moreover, intensive work was carried out aimed at developing the reproductive protocol (Migaud et al. 2004) and nutritional requirements (Henrotte et al. 2010a, b) of domesticated broodstock (see also Fontaine et al. 2015).

Most recently, significant efforts were devoted to the investigation of the domestication process (Douxfils et al. 2011a, b) and optimization of protocols of induction of gonadal development and spawning (Abdulfatah et al. 2011, 2013). However, despite of all these efforts, there are still no clear reproductive protocols, which in domesticated fish were scarcely studied. Together with the widely observed low and/or variable quality of eggs, this renders the entire production cycle of Eurasian perch in RAS of low efficiency, although already feasible. On the contrary, intensive research activities were conducted for the development of spawning protocols for wild (and/or pond-reared) Eurasian perch which were easily accessible and the reproductive outcome was not masked by the improper husbandry conditions. Therefore, the studies on the controlled reproduction of the wild specimens are always the first step toward the intensive aquaculture production as it allows to understand the mechanisms conditioning the effectiveness of controlled reproduction (Policar et al. 2008; Szczerbowski et al. 2009; Rónyai and Lengyel 2010; Żarski et al. 2011a, 2011b).

The popularity of percids is reflected also in the number of scientific and professional stakeholders working with these fish. Increasing interest in the culture of percid fish on European level resulted in the formation of a thematic group (European Percid Fish Culture) of the European Aquaculture Society in 2012. A book entitled "Biology and Culture of Percid Fishes" has been edited and published in 2015, also signifying the importance of this topic. The COST Action FA1205 AQUAGAMETE is also involved in the work on percids, in particular in topics related to gametogenesis and reproduction. Authors of the current volume are also active participants of this Action.

Controlled reproduction may be defined as a human intervention in the process of reproduction (Woynarovich and Horvath 1980) which takes into account a number of techniques leading to production of high quality offspring. These techniques usually include all the protocols aiming at induction of final gonadal maturation and

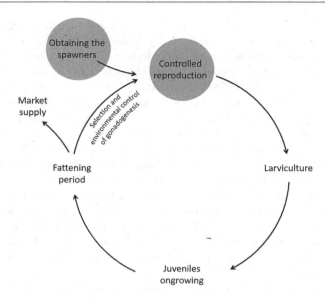

Fig. 1.3 Aspects undertaken within the present handbook (encircled in *blue*) from among the entire cycle of intensive aquaculture production process, with the wild (or pond-reared) fish used as a model

spawning as well as gamete management and incubation, up to hatching (Żarski et al. 2015b). The present Hatchery Manual gathers the currently existing knowledge on these very first steps of intensive aquaculture production related with obtaining the spawners from the wild (wild-like environments, such as earthen ponds) and controlled reproduction of Eurasian perch (Fig. 1.3). However, this handbook is not a typical literature review. Beside a briefly explained rationale and scientific background standing behind each of the steps of controlled reproduction, the entire monograph is extensively supplemented by specific practical recommendations which, to the best of our current knowledge, can be considered as the optimal ones allowing to perform successful reproduction of Eurasian perch. One thing that must be emphasized is that all described procedures and methods were developed with the use of wild and/or pond-reared fish as a model. This allowed to exclude the effect of improper environmental conditions as well as nutritional aspects that can have an influence on gamete quality as much as possible. Of course, the negative effect of global warming as well as some other environmental disturbances should not be excluded as a modulatory factors of the spawning effectiveness of the wild fish. Therefore, the possible comparison of the reproductive outcomes of different populations or the same population over the years should be always considered with high dose of criticism. Additionally, it must be highlighted that none of the described protocols considered the possibly different physiological reactions of domesticated fish (RAS-reared) to a particular steps of the reproductive protocol. This should be taken into the consideration before transferring the methods described here to commercial production of cultured stocks. Nonetheless, we hope that the handbook will be a valuable contribution allowing better understanding of most of the problems associated with the controlled reproduction of this species, get to know the specific methods developed and applied up to date, as well as to recognize the main bottlenecks of controlled reproduction. Despite the fact, that the protocols presented here should be first adapted to the cultured stocks with different level of domestication (Teletchea and Fontaine 2014) as well as different genetic background, we hope that the set of the information presented will allow further effective development of the intensive aquaculture of Eurasian perch.

The preparation of this Hatchery Manual was a part of the PERCAHATCH (E!8028) project funded under the auspices of the EUREKA funding scheme (for details go to: www.eurekanetwork.org). The project was initiated by a Szabolcsi Halászati Ltd. (represented by Mr. János Radóczi) in close collaboration with Szent Istvan University (Gödöllő, Hungary) and University of Warmia and Mazury (Olsztyn, Poland). The structure of the Consortium allowed effectively elaborate, evaluate and practically implement many of the protocols, which were described in the present handbook. In some cases the protocols are based on unpublished data, since before the accomplishment of the project many of the collected data were still being prepared for publication. Nevertheless, all procedures and recommendations described here were already practically evaluated thanks to the close collaboration with Szabolcsi Halaszati Ltd., which turned into a set of practical advices also incorporated in this monograph. However, although the protocols recommended were found to be the most effective among the tested ones, we additionally recommend to follow the described procedures critically taking into account that reproduction outcome is always a cumulative effect of all preceding events in the fish life history and that an effective and repeatable protocol must be improved continuously under local hatchery conditions.

References

Abdulfatah A, Fontaine P, Kestemont P, Gardeur J-N, Marie M (2011) Effects of photothermal kinetics and amplitude of photoperiod decrease on the induction of the reproduction cycle in female Eurasian perch Perca fluviatilis. Aquaculture 322–323:169–176. doi:10.1016/j.aquaculture.2011.09.002

Abdulfatah A, Fontaine P, Kestemont P, Milla S, Marie M (2013) Effects of the thermal threshold and the timing of temperature reduction on the initiation and course of oocyte development in cultured female of Eurasian perch Perca fluviatilis. Aquaculture 376–379:90–96. doi:10.1016/j.aquaculture.2012.11.010

Abi-Ayad SMEA, Kestemont P, Mélard C (2000) Dynamics of total lipids and fatty acids during embryogenesis and larval development of Eurasian perch (Perca fluviatilis). Fish Physiol Biochem 23:233–243. doi:10.1023/A:1007891922182

Baras E, Kestemont P, Mélard C (2003) Effect of stocking density on the dynamics of cannibalism in sibling larvae of Perca fluviatilis under controlled conditions. Aquaculture 219:241–255. doi:10.1016/S0044-8486(02)00349-6

Cuvier-Péres A, Jourdan S, Fontaine P, Kestemont P (2001) Effects of light intensity on animal husbandry and digestive enzyme activities in sea bass Dicentrachus labrax post-larvae. Aquaculture 202:317–328. doi:10.1016/S0044-8486(01)00781-5

Douxfils J, Mandiki SNM, Marotte G, Wang N, Silvestre F, Milla S, Henrotte E, Vandecan M, Rougeot C, Mélard C, Kestemont P (2011a) Does domestication process affect stress response in juvenile Eurasian perch Perca fluviatilis? Comp Biochem Physiol Part A Mol Integr Physiol 159:92–99. doi:10.1016/j.cbpa.2011.01.021

Douxfils J, Mathieu C, Mandiki SNM, Milla S, Henrotte E, Wang N, Vandecan M, Dieu M, Dauchot N, Pigneur LM, Li X, Rougeot C, Mélard C, Silvestre F, van Doninck K, Raes M, Kestemont P (2011b) Physiological and proteomic evidences that domestication process differentially modulates the immune status of juvenile Eurasian perch (Perca fluviatilis) under chronic confinement stress. Fish Shellfish Immunol 31:1113–1121. doi:10.1016/j.fsi.2011.10.001

Fiogbé ED, Kestemont P, Mélard C, Micha JC (1996) The effects of dietary crude protein on growth of the Eurasian perch Perca fluviatilis. Aquaculture 144:239–249. doi:10.1016/S0044-8486(96)01293-8

Fontaine P (2004) L'elevage de la perche commune, une voie de diversification pour l'aquaculture continentale. Prod Anim 17:189–193

Fontaine P, Tamazouzt L, Capdeville B (1996) Growth of the Eurasian perch (Perca fluviatilis L.) reared in floating cages and in water recirculated system: first results. J Appl Ichthyol 12:181–184

Fontaine P, Gardeur JN, Kestemont P, Georges A (1997) Influence of feeding level on growth, intraspecific weight variability and sexual growth dimorphism of Eurasian perch Perca fluviatilis L. reared in a recirculation system. Aquaculture 157:1–9

Fontaine P, Legendre M, Vandeputte M, Fostier A (2009) Domestication of new species and sustainable development in fish culture. Cah Agric 18:119–124. doi:10.1684/agr.2009.0293

Fontaine P, Wang N, Hermelink B (2015) Broodstock management and control of the reproductive cycle. In: Kestemont P, Dąbrowski K, Summerfelt RC (eds) Biology and culture of percid fishes. Springer Netherlands, Dordrecht, pp 103–122

Henrotte E, Kaspar V, Rodina M, Psenicka M, Linhart O, Kestemont P (2010a) Dietary n-3/n-6 ratio affects the biochemical composition of Eurasian perch (Perca fluviatilis) semen but not indicators of sperm quality. Aquac Res 41:e31–e38. doi:10.1111/j.1365-2109.2009.02452.x

Henrotte E, Mandiki RSNM, Prudencio AT, Vandecan M, Mélard C, Kestemont P (2010b) Egg and larval quality, and egg fatty acid composition of Eurasian perch breeders (Perca fluviatilis) fed different dietary DHA/EPA/AA ratios. Aquac Res 41:e53–e61. doi:10.1111/j.1365-2109.2009.02455.x

Jourdan S, Fontaine P, Boujard T, Vandeloise E, Gardeur J, Anthouard M, Kestemont P (2000) Influence of daylength on growth, heterogeneity, gonad development, sexual steroid and thyroid levels, and N and P budgets in Perca fluviatilis. Aquaculture 186:253–265. doi:10.1016/S0044-8486(99)00357-9

Kestemont P, Mélard C, Fiogbé E, Vlavonou R, Masson G (1996) Nutritional and animal husbandry aspects of rearing early life stages of Eurasian perch Perca fluviatilis. J Appl Ichthyol 12:157–165

Kouril J, Linhart O, Relot P (1997) Induced spawning of perch by means of a GnRH analogue. Aquac Int 5:375–377

Kucharczyk D, Kujawa R, Mamcarz A, Skrzypczak A, Wyszomirska E (1996) Induced spawning in perch, Perca fluviatilis L. using carp pituitary extract and HCG. Aquac Res 27:847–852. doi:10.1046/j.1365-2109.1996.t01-1-00802.x

Kucharczyk K, Mamcarz S, Wyszomirska E (1998) Induced spawning in perch, Perca fluviatilis L., using FSH + LH with pimozide or metoclopramide. Aquac Res 29:131–136. doi:10.1046/j.1365-2109.1998.00949.x

Mandiki SNM, Houbart M, Babiak I, Vandeloise E, Gardeur JN, Kestemont P (2004) Are sex steroids involved in the sexual growth dimorphism in Eurasian perch juveniles? Physiol Behav 80:603–609. doi:10.1016/j.physbeh.2003.10.016

Martins CIM, Eding EH, Verdegem MCJ, Heinsbroek LTN, Schneider O, Blancheton JP, D'Orbcastel ER, Verreth JAJ (2010) New developments in recirculating aquaculture systems in Europe: a perspective on environmental sustainability. Aquac Eng 43:83–93. doi:10.1016/j.aquaeng.2010.09.002

Mélard C, Baras E, Mary L, Kestemont P (1996) Relationships between stocking density, growth, cannibalism and survival rate in intensively cultured larvae and juveniles of perch (Perca fluviatilis). Ann Zool Fenn 33:643–651

Migaud H, Fontaine P, Sulistyo I, Kestemont P, Gardeur JN (2002) Induction of out-of-season spawning in Eurasian perch Perca fluviatilis: effects of rates of cooling and cooling durations on female gametogenesis and spawning. Aquaculture 205:253–267. doi:10.1016/S0044-8486(01)00675-5

Migaud H, Gardeur JN, Kestemont P, Fontaine P (2004) Off-season spawning of Eurasian perch Perca fluviatilis. Aquac Int 12:87–102. doi:10.1023/B:AQUI.0000017190.15074.6c

Nielsen R, Asche F, Nielsen M (2015) Restructuring European freshwater aquaculture from family-owned to large-scale firms – lessons from Danish aquaculture. Aquac Res n/a-n/a. doi:10.1111/are.12836

Overton JL, Toner D, Policar T, Kucharczyk D (2015) Commercial production: factors for success and limitations in European percid fish culture. In: Kestemont P, Dąbrowski K, Summerfelt RC (eds) Biology and culture of percid fishes. Springer Netherlands, Dordrecht, pp 881–890

Policar T, Kouril J, Stejskal V, Hamackova J (2008) Induced ovulation of perch (Perca fluviatilis L.) by preparations containing GnRHa with and without metoclopramide. Cybium 32:308

Rónyai A, Lengyel SA (2010) Effects of hormonal treatments on induced tank spawning of Eurasian perch (Perca fluviatilis L.). Aquac Res 41:e345–e347. doi:10.1111/j.1365-2109.2009.02465.x

Steenfeldt S, Fontaine P, Overton JL, Policar T, Toner D, Falahatkar B, Horváth Á, Ben KI, Hamza N, Mhetli M (2015) Current status of Eurasian percid fishes aquaculture. In: Kestemont P, Dąbrowski K, Summerfelt RC (eds) Biology and culture of percid fishes. Springer Netherlands, Dordrecht, pp 817–841

Sulistyo I, Fontaine P, Rinchard J, Gardeur JN, Migaud H, Capdeville B, Kestemont P (1998) Reproductive cycle and plasma levels of steroids in male Eurasian perch Perca fluviatilis. Aquat Living Resour 11:101–110. doi:10.1016/S0990-7440(00)00146-7

Szczerbowski A, Kucharczyk D, Mamcarz A, Łuczyński MJ, Targońska K, Kujawa R (2009) Artificial off-season spawning of Eurasian perch Perca fluviatilis L. Archive 17:95–98

Tamazouzt L, Chatain B, Fontaine P (2000) Tank wall colour and light level affect growth and survival of Eurasian perch larvae (Perca fluviatilis L.). Aquaculture 182:85–90. doi:10.1016/S0044-8486(99)00244-6

Teletchea F, Fontaine P (2014) Levels of domestication in fish: implications for the sustainable future of aquaculture. Fish Fish 15:181–195. doi:10.1111/faf.12006

Toner D (2015) The market for Eurasian perch. In: Kestemont P, Dąbrowski K, Summerfelt RC (eds) Biology and culture of percid fishes. Springer Netherlands, Dordrecht, pp 865–879

Wang N, Gardeur JN, Henrotte E, Marie M, Kestemont P, Fontaine P (2006) Determinism of the induction of the reproductive cycle in female Eurasian perch, Perca fluviatilis: identification of environmental cues and permissive factors. Aquaculture 261:706–714. doi:10.1016/j.aquaculture.2006.08.010

Woynarovich E, Horvath L (1980) The artificial propagation of warm-water finfishes – a manual for extension. Food and Agriculture Organization of the United Nations, Rome

Xu X, Kestemont P (2002) Lipid metabolism and FA composition in tissues of Eurasian perch Perca fluviatilis as influenced by dietary fats. Lipids 37:297–304

Żarski D, Bokor Z, Kotrik L, Urbanyi B, Horváth A, Targońska K, Krejszeff S, Palińska K, Kucharczyk D (2011a) A new classification of a preovulatory oocyte maturation stage suitable for the synchronization of ovulation in controlled reproduction of Eurasian perch Perca fluviatilis L. Reprod Biol 11:194–209. doi:10.1016/S1642-431X(12)60066-7

Żarski D, Palińska K, Targońska K, Bokor Z, Kotrik L, Krejszeff S, Kupren K, Horváth Á, Urbányi B, Kucharczyk D (2011b) Oocyte quality indicators in Eurasian perch, Perca fluviatilis L., during reproduction under controlled conditions. Aquaculture 313:84–91. doi:10.1016/j.aquaculture.2011.01.032

Żarski D, Horváth A, Held JA, Kucharczyk D (2015) Artificial reproduction of percid fishes. In: Kestemont P, Dąbrowski K, Summerfelt RC (eds) Biology and culture of percid fishes, 1st edn. Springer Netherlands, Dordrecht, pp 123–161

Harvest, Transport of Spawners, Prophylaxis

2

Harvesting of Spawners

There exist two ways of harvesting perch spawners. One – from natural bodies of water and the other – from earthen pond systems (e.g. carp farms, polyculture farms, dam reservoirs). The spawners can be harvested in carp farms either in the spring or in the autumn. Spawners harvested in the spring are to be immediately moved to a hatchery and the procedure of controlled spawning is to be commenced instantly. Spawners harvested in the autumn can be moved to wintering ponds and kept there until spring, or they can be moved to a hatchery and an 'advanced spawning' (up to several months before the spawning season; for details see: Chap. 11) procedure can be started. If the fish spend the winter in a wintering pond, the spawners should be provided with a proper amount of natural food (small-size fish like common roach or sunbleak).

It is best to harvest perch spawners from natural bodies of water in the spring. In this period the fish gather on the spawning grounds and there is a possibility to harvest a significant number of spawners in a short period of time. All types of fishing equipment can be used for catching fish in natural environment. However, it is not recommended to use set gill nets, as they harm the fish. Best results are achieved when fish traps are used or the fish are caught with the drag nets.

Regardless the place of origin, perch spawners should display proper liveliness, health, built and condition. When kept in water, they should take proper body position and naturally react to external stimuli. Health-wise, the fish should be characterized by the lack of external symptoms of disease, such as, thrush or parasites visible to the naked eye. They should have proper body, mouth, gills and fins structure. Apart from the above, they should be characterized with the lack of symptoms of anemia and no mechanical body damages, especially to the mouth.

© The Authors 2017
D. Żarski et al., *Controlled Reproduction of Wild Eurasian Perch*, SpringerBriefs in Environmental Science, DOI 10.1007/978-3-319-49376-3_2

Transporting of Spawners

It is recommended to use containers equipped with oxygenation systems for transporting spawners. The containers should be made of materials which have no harmful effect on live fish. Their edges and sidewalls should be smooth so as not to cause mechanical injuries to the skin of the fish. Compressed air tank is to be used for water oxygenation. Oxygenating systems should be equipped with smooth-surface hoses. Air nozzles should provide even oxygenation of the whole water volume. Auxiliary equipment (landing nets, buckets, tubs and so on) should be disinfected. The water used for fish transport cannot contain chlorine or ammonia. For transportation 'natural' water (from the pond or the lake – where the fish were kept before the transportation) or the tap water (if it is chlorine-free and after it is oxygenated) can be used. Water temperature differences between the containers were the fish were kept and the transport container should not exceed 2 °C. A small number of perch spawners can also be transported in polyethylene bags.

Perch spawners due for transportation must be fasted for at least 24 h. The fish due for transportation are best to be weighed along with the previously tared container partially filled with water. After weighing, the fish should be transferred to the transport container or a polyethylene bag. The transport container should be filled with water to $^1/_3$ of its capacity before loading and then to $^2/_3$ of its capacity after loading. Polyethylene bags should be filled with water to the half of their capacity, the fish should be loaded, the air removed from the bag, the oxygen pumped in and then the bag should be closed tightly. Loading should be performed in a continuous, smooth and delicate manner, so as not to hit the fish or cause injuries.

During transportation of the fish all rapid movements causing waves on the water or spilling over the edges of the container are to be avoided. It is also important that the temperature of the water in the fish container does not increase. If the time of the transportation is expected to be longer or the fish load is higher than presented on Table 2.1, it is necessary to exchange the water in the tank. To this end, the water is to be added slowly and evenly all over the surface of the container. After arriving at the destination, the unloading is to be carried out in accordance with the same rules as applied during the loading. Above all, the temperatures differences between the transport container and the new environment for the fish are to be equaled.

The number of the fish safe for transportation depends on the condition of the fish (e.g. caused by handling practices prior to transportation), but mainly on the water temperature and the time of the transport. The fish mass is to be chosen per water volume unit. Until present, no guidelines have been developed concerning the transportation of perch spawners. Therefore, in order to carry out the transportation of these fish, it is recommended to apply the guidelines concerning pikeperch or generally percid fish (Tables 2.1 and 2.2).

Table 2.1 Indicative conditions for 5–20 h transports of fish in tank with oxygen supply

Individual weight of fish (g)	Amount of fish (kg) in 1000 liter water at				
	0–5 °C	5–8 °C	8–10 °C	10–15 °C	15–20 °C
Under 100	50–100	40–80	30–60	24–48	20–40
100–200	100–125	80–100	60–75	48–60	40–50
200–500	125–175	100–140	75–105	60–84	50–70
500–1000	175–200	140–160	105–120	84–96	70–80
1000	250	200	150	120	100
1000–1700	275–288	220–230	165–173	132–138	110–115

Berka (1986)

Table 2.2 The amounts of fish in kg to be transported in 40–litre bags containing 20 litres of water and 20 litres of oxygen

Temperature (°C)	Duration of transport (in h)									
	5	10	15	20	25	30	35	40	45	50
5	1.8	1.8	1.8	1.8	1.8	1.8	1.6	1.4	1.3	1.2
10	1.8	1.8	1.8	1.3	1.0	0.93	0.8	0.7	0.62	0.056
15	1.8	1.8	1.8	1.3	1.0	0.93	0.8	0.7	0.62	0.56
20	1.8	1.8	1.4	1.0	0.9	0.75	0.64	0.56	0.5	0.45

Berka (1986)

Prophylaxis

The first activity after transporting the spawners to the hatchery should be inspection of the exterior of the fish. Perch spawners, especially after capture in open waters, are prone to abrasions and injuries. The extent of these injuries depends on the method of capture and the manner of handling the spawners. All types of injuries are to be disinfected as soon as possible, as they are a welcoming ground for infections which can lead to death. An aqueous solution of gentian violet is recommended for disinfection; it is commonly available in pharmacies and safe to use, as it is intended for application in humans. Its greatest advantage is easy administration; it should be applied to the abrasion and injured areas, similarly to its use for human skin abrasion and injury. One-time disinfection may not be effective; therefore the disinfection is to be repeated with every transfer of fish.

During the inspection of the fish exterior, attention must be paid to the health condition of the fish. All anomalies in their external appearance and a high number of parasites may exclude the fish from being breeders. Therefore, if any alarming symptoms are noticed, a veterinarian is to be contacted as soon as possible. The veterinarian should assess the health condition of the fish and supervise the treatment. It is not recommended to treat the fish without professional supervision, as the lack of knowledge concerning a given medicine or its improper application may lead to an adverse effect. For more information on parasitic infections in perch as well as the other diseases please see Rodger and Phelps (2015) and Behrmann-Godel and Brinker (2016).

In order to compensate for the negative effects of stress caused by the transport and transfer, 1–2 kilos of sodium chloride per 1 m^3 may be added to the spawning containers. It is to be administered first time immediately after transporting the fish into the hatchery. Subsequent applications should be carried out in 1-week intervals.

References

Behrmann-Godel J, Brinker A (2016) Biology and ecologyu of perch parasites. In: Couture P, Pyle G (eds) Biology of perch. CRC Press, Boca Raton, pp 193–229

Berka R (1986) The transport of live fish. a review. EIFAC technical paper 48. Food and Agriculture Organization of the United Nations, Rome, p 52

Rodger HD, Phelps NBD (2015) Percid fish health and disease. In: Kestemont P, Dąbrowski K, Summerfelt RC (eds) Biology and culture of percid fishes. Springer Netherlands, Dordrecht, pp 799–813

Hatchery Manipulation and Broodstock Selection

3

Theoretical Background

The successful reproduction is highly dependent on the overall condition of the spawners. It concerns, among others, the maturation (whether the fish is ready to spawn or not) as well as health status of the fish. The latter is especially important when wild and pond-reared fish are reproduced since the capture/harvesting procedure affects skin damages and altered immune response caused by high stress. This, in turn, creates favorable conditions for any kind of infections (especially fungal and bacterial) which may even lead to high losses in the broodstock. Even if a fish in poor condition (i.e. exhibiting some external damages, fin degradation, lack or high turbidity of mucus, opaque and whitish scales) might survive the spawning procedure, it should be expected with high probability, that the quality of gametes obtained from such a fish can be very low. Therefore, the selection of fish for the spawning procedure is one of the key steps during controlled reproduction. However, although the fish designated for spawning will be properly chosen there is still a number of manipulations, to which the fish will be exposed before the gametes will be acquired. And each handling procedure is a stress factor creating the risk of lowering the health status of the fish. That is why, knowledge of the basic principles of hatchery manipulations, together with some prophylaxis methods (described in greater details in Chap. 2) is of obvious importance from the commercial point of view.

For more detailed theoretical background of controlled reproduction of percid fishes we highly recommend to see also Żarski et al. (2015).

Conditions for Keeping the Spawners

Spawners of Eurasian perch (especially wild and pond-reared ones) are very sensitive to keeping in controlled conditions. In addition to frequent manipulations, the fish are also exposed to unnaturally high densities which is a stress factor, as well (Bly et al. 1997). Therefore, the conditions of keeping spawners for controlled

D. Żarski et al., *Controlled Reproduction of Wild Eurasian Perch*, SpringerBriefs in Environmental Science, DOI 10.1007/978-3-319-49376-3_3

reproduction should carefully be reconsidered. Fish, depending on the size, should be kept in tanks with a total volume between 300 l (for fish with an average weight of maximum 150 g) to maximum 1000 l. Larger tanks are not recommended since it is much more difficult to catch the fish with a net and the long 'chasing' of the fish prior to their removal is another stress factor. The tanks should have smooth surfaces without any submerged items in it. It is not recommended to use tanks where fish can find hiding places (cavities, submerged outflow pipes etc.) since fish exhibit a tendency to gather in such places which can lead to skin damages (by the submerged items as well as the fin spines of other individuals). Lighting should be distributed homogeneously as much as possible over the entire surface of the tank to prevent gathering of the fish in the darker parts of the tank. The light should not be too intensive and should not be higher than 300 lx on the surface of the tank. However, the most preferable is to keep the wild and pond-reared fish under constant dimness (below 50 lx). Especially because the photoperiod plays a minor role in the final oocyte maturation and spawning (for details see the Chap. 5) and in constant dimness fish receive less stimuli (behave much more calmly) which presumably limits the stress. The tanks should be supplied with mechanically and biologically filtered (ammonium- and nitrite-free) water at a constant temperature. The water flow should allow to exchange the entire water volume in the tank at least within 1 h. The temperature should be fully controlled. Directly in the holding tanks, it should be possible to maintain a constant temperature (± 0.2 °C) between 10 and 16 °C, whereas extreme values should be possible to reach within less than 6 h.

Water Quality Parameters

Recirculated aquatic systems should provide the perch spawners with proper environmental conditions. The following physical-chemical properties of the water are to be kept during keeping the fish:

- Water temperature from 5 to 20 °C.
- pH (7, 0 – 7, 5)
- Total ammonia nitrogen CA (<0, 0125 mg/dm^3)
- nitrite nitrogen NO_2-N (< 0,005 mg/dm^3)
- Oxygenation (at least 80 % saturation)
- Mechanical and biofiltration efficiency – general slurry < 1 mg/dm^3

Sex Recognition

During the final phases of controlled reproduction it may be beneficiary to separate the fish of both genders. This allows to avoid possible pheromonal interactions between males and females (Żarski 2012) and, in turn, adjust the spawning protocol to be more reliable and predictable. That is why the sex recognition is an important aspect of controlled reproduction. However, in the case of Eurasian perch there is no

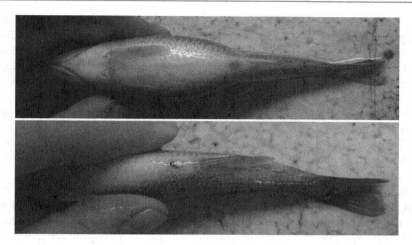

Fig. 3.1 Slightly distended belly of the female (at the *top*) and a male similar in size with a small but well noticeable droplet of sperm released from urogenital pore (appeared after gentle pressure of abdomen) (at the *bottom*) (Photo: Z. Bokor)

visible sexual dimorphism. This allows to distinguish males from females by the external features only during the spawning season, when females usually have a swollen abdomen and males usually release a small volume of sperm after a gentle pressure of the belly (Fig. 3.1). This latter feature may sometimes be useful in the case of 'advanced spawning' of wild and pond-reared fish since males sometimes release some sperm even during the winter (e.g. Alavi et al. 2010). However, in the case of 'advanced spawning' the possibility of sex recognition is still rather limited. This applies also to cultured fish kept under intensive culture conditions (in RAS) when the two sexes are often morphologically very similar. Nonetheless, sex recognition should be performed very carefully and very critically regardless the fish origin.

One of the easiest methods of verification of the sex in fish is catheterization (Ross 1984; Smith et al. 2014). For that purpose, a catheter should gently be inserted into the urogenital pore (in the present manual referred also to as 'genital papilla') of the fish and suction by a syringe attached to the catheter should be used to collect a small sample of the gonad, if possible (described in details in Chap. 4). The catheterization, if performed properly, is usually not harmful to the fish and can safely be applied several times before spawning (Żarski et al. 2011).

Hatchery Manipulations

During controlled reproduction fish are usually exposed to a number of stressors. This applies to frequent changes of the environment in which fish are kept (including photo-thermal regime, confinement in the holding tanks etc.) as well as all the necessary manipulations. Depending on the purpose of each handling procedure, fish can either only be moved between tanks (with the use of net or by hand) or subjected to specific manipulations such as catheterization, injection or stripping. It is important to realize, that regardless of the method of manipulation, each handling procedure is very stressful as well as causes mechanical removal of the mucus from the skin. This is especially the case during manipulations when the fish are in direct contact with, e.g. benchtops, cloths etc. Considering the fact that the mucus is the first line of defense of the fish (Shephard 1994; Woof and Mestecky 2005), special attention should be paid to avoiding complete mucus removal. Especially, that the perch is a species with relatively low mucus layer. Additionally, any kind of practices enhancing mucus production and stress mitigation (such as short-term baths in a NaCl solution) (Tacchi et al. 2015) should be considered as an important part of the handling protocols.

Manipulations may also have an indirect adverse effect on reproduction outcome. As a direct stress factor, handling (especially performed improperly or too frequently) may negatively affect the quality of gametes through hormonal responses of the fish organism to the stressor (Schreck et al. 2001; Barton 2002; Schreck 2010). In order to limit the negative effects of manipulations, it is recommended to use anesthetics before each manipulation (Kristan et al. 2012, 2014; Gomułka et al. 2015b). However, the application of anesthetics should not only be considered from the perspective of stress reduction but also as an indispensable procedure before each manipulation that can cause pain to the fish (e.g. hormonal injection) (Gomułka et al. 2015b). It must be emphasized, that all anesthetics are also substances that provoke a stress reaction in the fish affecting hematological indices and alterations in osmoregulation (Kristan et al. 2012; Gomułka et al. 2014, 2015a), although this varies according to the type of chemical, potency and species. Therefore, the choice of a particular anesthetic should be species-dependent (Gomułka et al. 2015b) as well as it should consider the national and/or regional law regulations.

Basic Information on Anesthesia

Anesthesia in fish is usually carried out by immersion. Practically, the fish are transferred from the holding tank into a smaller, aerated container/tank containing an anesthetic solution. Anesthetics usually produce four different levels of anesthesia. This includes (e.g. Coyle et al. 2004):

Sedation – the breathing and motion of the fish are reduced
Anesthesia – partial loos of equilibrium, still maintaining reaction to touch stimuli
Surgical anesthesia – fish completely lose equilibrium and do not react to touch stimuli
Death – breathing and heart beat stops leading to death.

Anesthesia levels reached depend on the dose and exposure time. Generally, the type and dose of the anesthetic should be considered that results in the shortest period between the exposure of fish to anesthetic bath and surgical anesthesia. This will allow shortening the period when the fish can harm themselves during struggling and facilitate the hatchery work when many fish are intended for anesthesia. However, Gilderhus and Marking (1987) indicates that the induction time (up to surgical anesthesia level) should not be shorter than 3 min. Shorter induction times may indicate too high doses that lead to reaching the fourth level (death) of anesthesia too fast. On the other hand, too long induction times will probably not allow to reach the third level of anesthesia properly and the fish (although immobilized) may still react to touch stimuli which is unacceptable during hatchery manipulations. Gilderhus and Marking (1987) suggested also that a secure dose should allow 15 min exposure of the fish to the anesthetic (after the fish reached surgical anesthesia level) after which fish should fully awaken within approximately 10 min in anesthetic-free water. Although, these theoretical assumptions should be taken into account while choosing the best anesthetic type and dose, each anesthesia operation should be preceded by initial observation how the fish react to particular anesthesia conditions (in fish with a particular physiological status, condition at particular water temperature) and modifications should immediately be applied in case of any worrying signals. In similar situations, the experience of the personnel may be invaluable for the success of the entire operation.

Practical Tips

Anesthesia

Necessary items (Fig. 3.2):

- Anesthetic (it is recommended to use MS-222 or other agent containing tricaine methanesulfonate).
- Tank for anesthesia (for perch the minimum size is a tank of about 20 l provided that 20 cm of depth will be guaranteed).
- Tank for a short-term bath in a salt solution (saline bath) with a salinity of 5 ppm.
- A recovery tank (similar or bigger than the tank for anesthesia).
- Aeration system (for providing gentle aeration in all tanks, including the tank with anesthetic; see Stoskopf and Posner 2008).

Important
Dose of MS-222 for Eurasian perch should amount to 150 mg l^{-1} (of pure active substance).

The procedure:

1. Prepare the anesthetic solution in the anesthesia tank.
2. Pour the anesthetic-free water into the recovery tank.
3. Provide gentle aeration in both tanks.
4. Transfer the fish gently from the holding tank into the anesthetic bath.

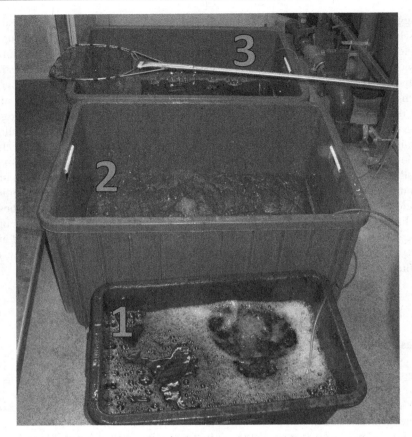

Fig. 3.2 Spot prepåred for manipulation of the Eurasian perch spawners (*1* – anesthetic bath, *2* – recovery tank, *3* – tank for short term saline bath). During manipulation fish can be easily transferred from one bath to another (Photo: S. Krejszeff)

5. After the fish reach the level of surgical anesthesia, gently remove the fish and carefully perform all nece.ssary manipulations.
6. Place the fish in the recovery tank.
7. After the fish fully recover (regain equilibrium and stabilize breathing movements), transfer the fish into the holding tank.

Important
1. Temperature of the anesthetic bath as well as the water in the recovery tank should be the same as in the holding tank from which the fish are taken.
2. The number of fish anesthetized simultaneously should be adjusted to the speed of the manipulations that have to be performed on fully anesthetized fish taking into account that the exposure time should not exceed 10 min (maximum 15).

(continued)

3. Do not put unrecovered fish back into the holding tank in order to make sure that all fish survived anesthesia.
4. After the recovery and before the transfer of the fish into the holding tank it is recommended to apply a short-term bath in a salt solution (for details see Chap. 2).

Sex Recognition

The necessary items:

- Catheter (the details on the catheter recommended are provided in Chap. 4).
- Syringe, 10 ml volume.
- Cloth (e.g. soft towel or cotton diaper).
- Paper towel.
- 90 % ethanol (for disinfection of the catheter).
- Anesthetic solution (e.g. solution of MS-222 at a dose of 150 mg l^{-1}).
- Manipulation bench (benchtop).

The procedure:

1. Anesthetize the fish (until the fish reach the level of surgical anesthesia).
2. Place the fish on the benchtop (onto a wet cloth) – belly up position.
3. Gently press the abdominal part of the body sliding the fingers on two sides of the belly from the ventral fins toward the urogenital pore.
4. Males can be recognized by a small droplet of sperm appearing in the urogenital pore (genital papilla – is the second pore from the head side, as shown on Fig. 4.7a).
5. If the sperm is not obvious, insert the catheter (connected to a syringe) into the urogenital pore of the fish into a depth allowing to maximally reaching the ventral fins of the fish.
6. The males should be identified if a small sample of sperm is drawn into the catheter, whereas in the case of females small sample of oocytes would appear in the catheter lumen (Fig. 3.3).
7. If in the particular fish it is not possible to acquire neither sperm nor oocytes, the individual should be considered as immature; in this case fish should be labeled (allowing further identification among the group of the fish – e.g. with a PIT tag) and should be kept for about 1–2 weeks with males (depending on the temperature) after which the procedure of sex recognition should be repeated.
8. Females should be placed in a separate holding tank(s), whereas males together with 'immature' individuals in the other one(s).

Fig. 3.3 Oocytes noticeable in the catheter lumen during the evaluation of the sex of the fish (Photo: Z. Bokor)

Practical Advice

In already mature fish, which are close to spawning (in wild fish) or were properly prepared for the 'advanced spawning' (for details see the Chap. 11), spermiating males usually herald the beginning of the maturation process. Therefore, if the males are not spermiating and it is not possible to catheterize most of the fish (among the representative group), it is recommended to keep all fish (males and females) together until sex determination in most fish is possible. During that period, temperatures between 10 and 12 °C are recommended and constant dimness (not more than 50 lx – in case of wild and pond-reared fish) or variable light conditions (between 10 and 14 h of light period per day) should be applied. At these temperatures the fish can be checked every 4–7 days (depending on the broodstock).

References

Alavi SMH, Rodina M, Hatef A, Stejskal V, Policar T, Hamáčková J, Linhart O (2010) Sperm motility and monthly variations of semen characteristics in *Perca fluviatilis* (Teleostei: Percidae). Czech J Anim Sci 55:174–182

Barton BA (2002) Stress in fishes: a diversity of responses with particular reference to changes in circulating corticosteroids. Integr Comp Biol 42:517–525. doi:10.1093/icb/42.3.517

Bly JE, Quiniou SM, Clem LW (1997) Environmental effects on fish immune mechanisms. Dev Biol Stand 90:33–43

Coyle SD, Durborow RM, Tidwell JH (2004) Anesthetics in Aquaculture. South Reg Aquac Cent:1–6

Gilderhus PA, Marking LL (1987) Comparative efficacy of 16 anesthetic chemicals on rainbow trout. North Am J Fish Manag 7:288–292. doi:10.1577/1548-8659(1987)72.0.CO

Gomułka P, Wlasow T, Szczepkowski M, Misiewicz L, Ziomek E (2014) The effect of propofol anaesthesia on haematological and biochemical blood profile of European whitefish. Turk J Fish Aquat Sci 14:331–337. doi:10.4194/1303-2712-v14_2_04

Gomułka P, Dągowski J, Własow T, Szczepkowski M, Czerniak E, Ziomek E, Szczerbowski A, Łuczyński M, Szkudlarek M (2015a) Haematological and biochemical blood profile in Russian sturgeon following propofol and eugenol anaesthesia. Turk J Fish Aquat Sci 15:13–17. doi:10.4194/1303-2712-v15_1_02

Gomułka P, Fornal E, Berecka B, Szmagara A; Ziomek E (2015b) Pharmacokinetics of propofol in rainbow trout following bath exposure. Pol J Vet Sci 18:147–152. doi:10.1515/pjvs-2015-0019

Kristan J, Stara A, Turek J, Policar T, Velisek J (2012) Comparison of the effects of four anaesthetics on haematological and blood biochemical profiles in pikeperch (Sander lucioperca L.). Neuroendocrinol Lett 33:66–71

Kristan J, Stara A, Polgesek M, Drasovean A, Kolarova J, Priborsky J, Blecha M, Svacina P, Policar T, Velisek J (2014) Efficacy of different anaesthetics for pikeperch (Sander lucioperca L) in relation to water temperature. Neuroendocrinology Letters. Maghira and Maas Publications, In, pp. 81–85

Ross RM (1984) Catheterization: a non-harmful method of sex identification for sexually monomorphic fishes. Prog Fish Cult 46:151–152. doi:10.1577/1548-8640(1984)46<151:C>2.0.CO;2

Schreck CB (2010) Stress and fish reproduction: the roles of allostasis and hormesis. Gen Comp Endocrinol 165:549–556. doi:10.1016/j.ygcen.2009.07.004

Schreck CB, Contreras-Sanchez W, Fitzpatrick MS (2001) Effects of stress on fish reproduction, gamete quality, and progeny. Aquaculture 197:3–24

Shephard KL (1994) Functions for fish mucus. Rev Fish Biol Fish 4:401–429. doi:10.1007/BF00042888

Smith GH, Murie DJ, Parkyn DC (2014) Nonlethal sex determination of the greater amberjack, with direct application to sex ratio analysis of the Gulf of Mexico stock. Mar Coast Fish 6:200–210. doi:10.1080/19425120.2014.927403

Stoskopf M, Posner LP (2008) Anesthesia and restraint of laboratory fish. In: RE F, Brown MJ, Danneman PJ, Karas AZ (eds) Anesthesia and analgesia in laboratory animals. Elsevier, San Diego, pp 519–534

Tacchi L, Lowrey L, Musharrafieh R, Crossey K, Larragoite ET, Salinas I (2015) Effects of transportation stress and addition of salt to transport water on the skin mucosal homeostasis of rainbow trout (Oncorhynchus mykiss). Aquaculture 435:120–127. doi:10.1016/j.aquaculture.2014.09.027

Woof JM, Mestecky J (2005) Mucosal immunoglobulins. Immunol Rev 206:64–82. doi:10.1111/j.0105-2896.2005.00290.x

Żarski D (2012) First evidence of pheromonal stimulation of maturation in Eurasian perch, Perca fluviatilis L., females. Turk J Fish Aquat Sci 12:771–776

Żarski D, Bokor Z, Kotrik L, Urbanyi B, Horváth A, Targońska K, Krejszeff S, Palińska K, Kucharczyk D (2011) A new classification of a preovulatory oocyte maturation stage suitable for the synchronization of ovulation in controlled reproduction of Eurasian perch Perca fluviatilis L. Reprod Biol 11:194–209. doi:10.1016/S1642-431X(12)60066-7

Żarski D, Horváth A, Held JA, Kucharczyk D (2015) Artificial reproduction of percid fishes. In: Kestemont P, Dąbrowski K, Summerfelt RC (eds) Biology and culture of percid fishes, 1st edn. Springer Netherlands, Dordrecht, pp 123–161

Determination of Maturity Stages of Oocytes

4

Theoretical Background

In fully mature Eurasian perch, being an iteroparous species (reproducing more than once during their lifetime) with a synchronous pattern of oocyte development (specific to a single-batch spawners), three general phases of the ovarian cycle may be distinguished: primary growth (up to the cortical alveoli stage), secondary growth (covering the early and late vitellogenesis) and final oocyte maturation (Khan and Thomas 1999; Fontaine et al. 2015). In perch, the beginning of vitellogenesis was usually observed at the end of July and the beginning of August (Le Cren 1951; Długosz 1986). The beginning of the migration of the germinal vesicle (GV) is considered as the end of the secondary growth phase (end of the vitellogenic period) (Żarski et al. 2012b). In the practice of controlled reproduction, all procedures are conducted when the process of final oocyte maturation (FOM) starts. For more details see also Fontaine et al. (2015) and Żarski et al. (2015). Therefore, in this manual the emphasis will be put on FOM and all the accompanying events.

FOM is a process during which a number of cellular changes occur in oocytes, leading to ovulation of fertilizable eggs. In perch this includes the GV migration from the center of oocyte to the cell periphery (towards the animal pole of the oocyte), coalescence of the oil droplets into a single large droplet, germinal vesicle break down (GVBD) and a hydration process accompanied by yolk homogenization (Patiño and Sullivan 2002; Nagahama and Yamashita 2008; Lubzens et al. 2010; Żarski et al. 2011a, b). In perch kept under controlled conditions this process (depending on the temperature regime) may last more than 30 days (Żarski 2012). From the perspective of commercial production the relatively long duration of this process is an unwanted and very problematic aspect. Even fish from the same population (stock) mature unevenly, causing a significant desynchronization of the spawning act. This, in turn, has a direct effect on practical aspects of the production process, as farmers have to consider acquiring small portions of eggs and consequently larvae hatching for a relatively long period. This usually creates the necessity of separate rearing of many small batches of larvae, otherwise the rearing

© The Authors 2017
D. Żarski et al., *Controlled Reproduction of Wild Eurasian Perch*, SpringerBriefs in Environmental Science, DOI 10.1007/978-3-319-49376-3_4

effectiveness may seriously be reduced by cannibalism as one of the the problematic aspects of perch larviculture (Mélard et al. 1996; Baras et al. 2003). Therefore for commercial purposes, hormonal stimulation is usually applied in order to speed up the FOM process and allow higher spawning synchronization (Kucharczyk et al. 1996, 1998; Żarski et al. 2011a).

Prediction of the moment of ovulation is a serious obstacle in controlled reproduction of percids since the fish do not mature simultaneously (Żarski et al. 2011a, 2012b). However, latency time between the administration of the hormonal agent and ovulation is strictly dependent on the spawning agent used (type and dose of hormone), maturation stage of fish (i.e. at which stage of FOM the particular female is) at the moment of stimulation and the applied thermal regime. Thus, maintenance of a constant temperature and precisely recognized maturation stages of females significantly facilitate to predict the moment of ovulation during induced spawning.

Maturation Stages of Oocytes

For controlled reproduction of perch it is advised to use a 6-stage classification of the maturation status of females which is based on the morphology (noticeable, after cytoplasm clarification, intracellular features) of preovulatory oocytes (as proposed by Żarski et al. 2011a).

Different levels of FOM are characterized by particular stages of maturation in Eurasian perch. The morphological features (as described by Żarski et al. 2011a; presented on Fig. 4.1), observed after the clarification of the cytoplasm (procedure of cytoplasm clarification described below), include:

Stage I – the germinal vesicle (GV; dark spot inside the oocyte) is in the central position or slightly moved toward the oocyte periphery (depending on the view angle), the cytoplasm is not transparent and contains fine granules in almost the entire area of the oocyte;

Stage II – the GV located in the center or slightly moved toward the edge is clearly surrounded by well visible oil droplets;

Stage III – the GV is clearly moved toward the animal pole of the oocyte and formation of large oil droplet(s) can easily be noticed, the forming oil droplet is similar in size or smaller than the GV;

Stage IV – the GV is clearly moved toward the oocyte periphery and a large (about 1/3 of the oocyte diameter) oil droplet may be noticed, however some smaller droplets are still present;

Stage V – the GV is clearly at the oocyte periphery and only one large oil droplet is visible;

Stage VI – the GV is not present anymore (GV breakdown is completed) and only one single oil droplet can be noticed. In this stage, freshly collected oocyte sample (before clarification of the cytoplasm) are already transparent and after

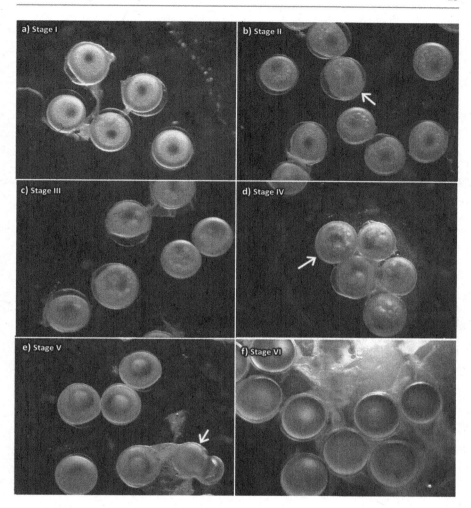

Fig. 4.1 Images of the preovulatory oocytes (taken under a stereoscopic microscope) representing different stages (from I to VI) of oocytes of Eurasian perch (fish originated from Sasek Wielki lake, Poland, oocytes catheterized during the 'advanced spawning' procedure, average weight 467±122 g). On image 'b' and 'd', an *arrow* indicates the oocyte which can be considered n stage III, whereas on image 'e', and *arrow* shows the oocyte which was damaged by catheterization (Photo: D. Żarski). The latency time (between the hormonal injection and ovulation) is specified in Table 6.2

immersion in Serra's solution they become slightly darkish (the clarification is not as efficient as in stages I-V), although the oil droplet is easily noticed.

> **Important**
> Determination of the position of the GV is strictly dependent on the orientation of the oocyte in relation to angle of view. Also, the location of the GV can be altered by changes in the internal structure of the oocyte caused by catheterization and the resulting pressure. In order to correctly determine the position of the GV it is required to roll and tilt each oocyte during microscopic observation.

The identification of maturation status of females based on the evaluation of the oocyte maturation stages is the only, so far, method allowing clear identification of the females at being at different maturation stages. It stems from the fact that despite the body weight of the female is considerably changing together with the advancement in maturation (from stage I to stage VI about 11 % of the weight increment can be observed; Fig. 4.2), as well as the gonadosomatic index (GSI) or hepatosomatic index (HSI) of the fish is changing dynamically during the FOM process (Fig. 4.3), the condition factor remains stable (Fig. 4.4). This indicates, that on the base of the external features as well as the morphometric analysis the maturation stage can not be predicted.

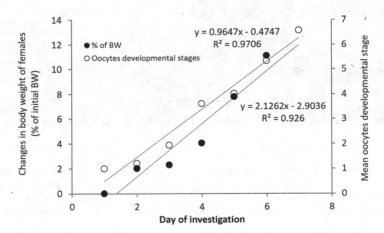

Fig. 4.2 Relationship between mean female body weight (BW) and oocyte maturity stage during the experiment on the determination the stage of oocyte maturity. Ovulation was induced outside the spawning season. Day 1 is a day of human chorionic gonadotropin (hCG; 500 IU/kg of body weight) injection. Initial BW corresponds to female weight at the time of hCG injection (According to Żarski et al. 2011a)

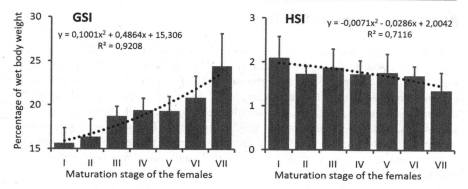

Fig. 4.3 Changes of the gonadosomatic (GSI; according to (Żarski et al. 2012b) and hepatoso-matic indices (HIS; Żarski D., unpublished) during the final oocyte maturation process in Eurasian perch. Indices for maturation stage VII (eggs were ovulated) were calculated taking into account weight of ovulated eggs (collected by stripping)

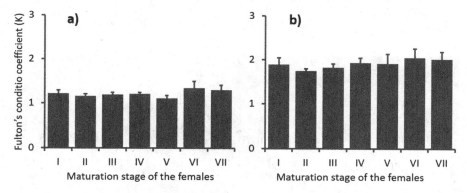

Fig. 4.4 Fulton's condition coefficient [calculated as follows: $K = 100 \, W \, L^{-3}$, where W is the body weight (g) and L is the length of the fish (cm)], calculated for total length (**a**) or for standard length (**b**). Indices for maturation stage VII (eggs were ovulated) were calculated taking into account weight of ovulated eggs (collected by stripping)

Specific Cases of the FOM Disruption

In some cases the process of FOM may be disrupted (by some independent factors, such as stress). This usually leads to anomalous changes in the oocytes during FOM. In these cases the developmental competence of ovulated eggs is impaired which is usually reflected in lower fertilization rate and/or developmental abnor-malities of the embryos and larvae. A very good example of the disrupted ovulation process is the occurrence of fragmented oil droplets in ovulated eggs, which are indicators of reduced egg quality (Żarski et al. 2011b). Although, egg quality can usually be determined after ovulation, in some cases these anomalies can be recog-nized before ovulation.

Anomalies Before the GVBD

Indication of any kind of anomalies during the early stages of FOM (before the GVBD) is very difficult. At that stage it is very difficult to say whether the morphological features observed stem from the slow maturation process or maybe the process is simply disturbed. The latter can clearly be identified only in the case when the oocytes sampled from the same female are characterized by divergent maturation stages, and these stages are not neighboring ones (Fig. 4.5). In this case it is recommended to remove the female from the further process of the controlled reproduction as the eggs ovulated by this female will most probably be characterized by highly reduced egg quality.

> **Important**
> The neighboring stages observed in the same sample of oocyte should be considered as a normal situation as oocytes very often do not mature perfectly simultaneously.

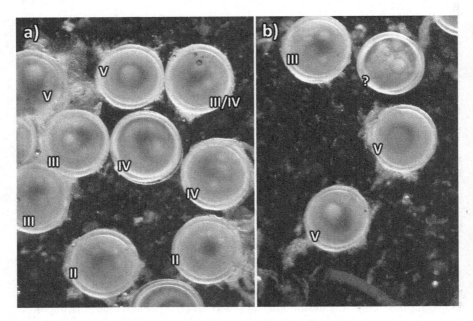

Fig. 4.5 Examples of oocytes from two different females (**a** and **b**) whose oocyte samples represented different maturation stages. Both fish originated from Sasek Wielki lake (Poland). Oocytes catheterized during the 'advanced spawning' procedure. Roman numerals indicate the maturation stage of the particular oocyte (according to the classification proposed by Żarski et al. 2011a). The question mark indicates an unclassified oocyte (this oocyte does not exhibit any typical damages caused by catheterization that would allow to speculate its 'abnormal' character) (photo: D. Żarski)

Anomalies After the GVBD

Anomalies of the FOM process in preovulatory oocytes can easily be recognized just after GVBD. Generally, the GVBD announces the completion of FOM and oocytes are just before the ovulation. Therefore, any abnormal cellular features of the oocytes at this stage clearly indicate that the FOM process was seriously disrupted and with the highest probability heralds highly reduced egg quality.

After the GVBD morphological features can be recognized before (as the eggs are usually already transparent) and after immersion into Serra's solution. In case of non-treated oocytes, two main indicators can be distinguished. First is a fragmented oil droplet and the second is internally damaged yolk and other abnormalities (Fig. 4.6a, c, and e). After the immersion of oocytes, fragmented oil droplets as well as internal cell damages and abnormalities can still be observed (Fig. 4.6b, d, and f). Actually, after the GVBD the immersion of the oocytes is usually not necessary anymore, however sometimes it helps to make sure that the germinal vesicle actually is not present anymore.

Practical Evaluation of the Maturity Stage of the Females

The necessary items:

- Catheter – among the available catheters it is recommended to use elastic (made of PVC with a Shore Hardness of approx. A80) infant feeding tubes (for enteral feeding of small babies and infants), which are characterized by the following features:
- Size: CH-06 (2 mm diameter)
- Length: 400 mm
- Round, sealed end with two openings on the surface near the end
- Equipped with a connector to a standard syringe;
- Syringe, 10 ml volume;
- Numbered Petri dishes (either plastic or glass);
- Clarifying Serra's solution (composition: 70-96 % ethanol, 36-40 % formaldehyde, 95 % glacial acetic acid in the proportion of 6:3:1);
- Numbered plastic buckets (with a minimum volume of 12 l) for a short-term storage of fish (the more buckets the more fish can be evaluated at the same time);
- Paper towel;
- 90 % ethanol (for disinfection of the catheter);
- Anesthetic solution (e.g. solution of MS-222 at a dose of 150 mg l^{-1});
- Manipulation table (benchtop).

Procedure of evaluation:

- Set the buckets (for a short-term storage of fish) into an order and fill them with a water (the same water in which the fish are kept).
- Prior to catheterization every fish must be anaesthetized.

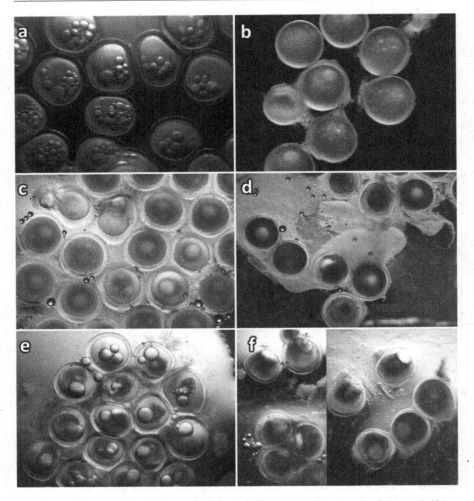

Fig. 4.6 Images of oocyte samples before (**a, c, e**) and after (**b, d, f**) immersion into the clarifying solution. Images (**a**) and (**b**) present an oocyte sample after GVBD where fragmented oil droplets can be observed. Images (**c**) and (**d**) present oocytes in different maturational stages (some were after GVBD) and with some internal damages noticeable after the immersion into the clarifying solution. Images (**e**) and (**f**) present oocytes with clearly visible internal damages and/or yolk fragmentation (Photo: D. Żarski)

- After the first fish is anesthetized, remove it from the anesthetic solution, place on the benchtop and insert the catheter gently into the genital papilla – approx. 2–3 cm (depending on the fish size) (Fig. 4.7b and c).
- By pulling the piston of the syringe connected to the catheter, draw an oocyte sample (~30–50 oocytes) into the catheter.
- Put the fish into the first bucket for recovery.
- Disconnect the catheter from the syringe and by applying gentle reverse pressure carefully blow out the oocyte sample into the first Petri dish.

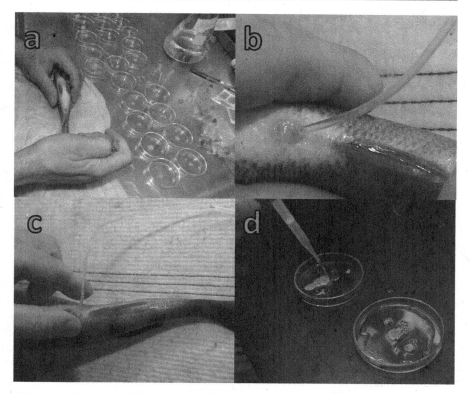

Fig. 4.7 Overview of the catheterization procedure: (**a**) – the benchtop prepared for catheterization of numerous fish (each Petri dish serves for the oocyte sample of a separate female) with a perch female being catheterized, (**b**) – proper placement of the catheter in the genital papilla, (**c**) – an oocyte sample drawn into a catheter, (**d**) – addition of the clarifying solution to an oocyte sample (Photo: S. Krejszeff)

- Immediately after pour Serra's solution onto the sample of oocytes, gently stir it and cover the Petri dish (Fig. 4.7d).
- Repeat steps 3–7 taking into account to assign the number of the Petri dish to a particular female.
- The samples should be evaluated under a stereoscopic microscope according to the classification given above.

Important
1. The oocytes usually do not mature perfectly simultaneously. Therefore, the criterion of classification of a given female to a particular stage should always be the majority (over 50 %) of the oocytes representing that particular stage.
2. Before oocyte sampling pay attention to keep catheter dry and possibly sterile.
3. During anesthesia comply with the rules described in Chap. 3.

References

Baras E, Kestemont P, Mélard C (2003) Effect of stocking density on the dynamics of cannibalism in sibling larvae of *Perca fluviatilis* under controlled conditions. Aquaculture 219:241–255. doi:10.1016/S0044-8486(02)00349-6

Długosz M (1986) Oogeneza i cykl rocznego rozwoju gonad wybranych gatunków ryb w zbiornikach o odmiennych warunkach termicznych. Acta Acad Agric Techn Olst, Prot Aquarum Piscat 14:1–68

Fontaine P, Wang N, Hermelink B (2015) Broodstock management and control of the reproductive cycle. In: Kestemont P, Dąbrowski K, Summerfelt RC (eds) Biology and culture of percid fishes. Springer Netherlands, Dordrecht, pp 103–122

Khan IA, Thomas P (1999) Ovarian cycle, Teleost Fish. In: Knobil E, Neill JD (eds) Encyclopedia of reproduction. Academic Press, San Diego, pp 552–564

Kucharczyk D, Kujawa R, Mamcarz A, Skrzypczak A, Wyszomirska E (1996) Induced spawning in perch, *Perca fluviatilis* L. using carp pituitary extract and HCG. Aquac Res 27:847–852. doi:10.1046/j.1365-2109.1996.t01-1-00802.x

Kucharczyk K, Mamcarz S, Wyszomirska E (1998) Induced spawning in perch, *Perca fluviatilis* L., using FSH + LH with pimozide or metoclopramide. Aquac Res 29:131–136. doi:10.1046/j.1365-2109.1998.00949.x

Le Cren ED (1951) The length-weight relationship and seasonal cycle in gonad weight and condition in the perch (*Perca fluviatilis*). J Anim Ecol 20:201–219. doi: masse poids taille methodologie

Lubzens E, Young G, Bobe J, Cerdà J (2010) Oogenesis in teleosts: how fish eggs are formed. Gen Comp Endocrinol 165:367–389. doi:10.1016/j.ygcen.2009.05.022

Mélard C, Baras E, Mary L, Kestemont P (1996) Relationships between stocking density, growth, cannibalism and survival rate in intensively cultured larvae and juveniles of perch (*Perca fluviatilis*). Ann Zool Fenn 33:643–651

Nagahama Y, Yamashita M (2008) Regulation of oocyte maturation in fish. Develop Growth Differ 50:S195–S219. doi:10.1111/j.1440-169X.2008.01019.x

Patiño R, Sullivan CV (2002) Ovarian follicle growth, maturation, and ovulation in teleost fish. Fish Physiol Biochem 26:57–70. doi:10.1023/A:1023311613987

Żarski D (2012) First evidence of pheromonal stimulation of maturation in Eurasian perch, *Perca fluviatilis* L., females. Turk J Fish Aquat Sci 12:771–776

Żarski D, Bokor Z, Kotrik L, Urbanyi B, Horváth A, Targońska K, Krejzeff S, Palińska K, Kucharczyk D (2011a) A new classification of a preovulatory oocyte maturation stage suitable for the synchronization of ovulation in controlled reproduction of Eurasian perch *Perca fluviatilis* L. Reprod Biol 11:194–209. doi:10.1016/S1642-431X(12)60066-7

Żarski D, Palińska K, Targońska K, Bokor Z, Kotrik L, Krejzeff S, Kupren K, Horváth Á, Urbányi B, Kucharczyk D (2011b) Oocyte quality indicators in Eurasian perch, *Perca fluviatilis* L., during reproduction under controlled conditions. Aquaculture 313:84–91. doi:10.1016/j.aquaculture.2011.01.032

Żarski D, Krejzeff S, Horváth Á, Bokor Z, Palińska K, Szentes K, Łuczyńska J, Targońska K, Kupren K, Urbányi B, Kucharczyk D (2012a) Dynamics of composition and morphology in oocytes of Eurasian perch, *Perca fluviatilis* L., during induced spawning. Aquaculture 364–365:103–110. doi:10.1016/j.aquaculture.2012.07.030

Żarski D, Kucharczyk D, Targońska K, Palińska K, Kupren K, Fontaine P, Kestemont P (2012b) A new classification of pre-ovulatory oocyte maturation stages in pikeperch, *Sander lucioperca* (L.), and its application during artificial reproduction. Aquac Res 43:713–721. doi:10.1111/j.1365-2109.2011.02879.x

Żarski D, Horváth A, Held JA, Kucharczyk D (2015) Artificial reproduction of percid fishes. In: Kestemont P, Dąbrowski K, Summerfelt RC (eds) Biology and culture of percid fishes, 1st edn. Springer Netherlands, Dordrecht, pp 123–161

Stimulation of Ovulation and Spermiation

5

Rationale of the Application of Hormonal Stimulation

In case of many farmed fish species, it is not possible to obtain high quality gametes without hormonal induction of the final phases of gamete maturation (Donaldson 1996; Mañanós et al. 2008). In the case of percids, however, it was already proven that photo-thermal manipulation is sufficient to promote the final stages of gamete maturation and consequently obtain satisfying spawning results (Müller-Belecke and Zienert 2008). However, in this case spawning is usually prolonged and may last even a few weeks (very often around 1 month). It may thus create serious negative consequences during commercial production. Prolonged spawning may result in excessive size differentiation of the larvae which leads to cannibalism (Kestemont et al. 2003; Baras et al. 2003) which, in turn, increase the labor intensity of the rearing procedure. Therefore, in the case of domesticated broodstock, hormonal stimulation is a useful tool that allows synchronization of spawning and shortening the spawning period.

In case of wild and pond-reared fish, it was also reported that hormonal stimulation was not absolutely necessary to obtain gametes (Rónyai and Lengyel 2010). Nonetheless, successful spawning without hormonal stimulation is strictly dependent on the maturation stage and condition of the fish. Generally, fish at stage VI would ovulate without any other treatment if kept in preferable spawning conditions (12–16 °C, photoperiod 12L:12D), although the moment of ovulation is unpredictable. There is also high chance that the fish, even if their oocytes are at stage VI of maturation, will not ovulate that can most probably be associated with the level of stress. Therefore, in case of wild and pond-reared fish hormonal stimulation is highly recommended even in fish that are at the most advanced stages of maturation. Apart from spawning synchronization it allows to predict the moment of ovulation, which is an indispensable condition allowing collection of 'dry' eggs for the further steps of controlled reproduction.

© The Authors 2017
D. Żarski et al., *Controlled Reproduction of Wild Eurasian Perch*, SpringerBriefs in Environmental Science, DOI 10.1007/978-3-319-49376-3_5

Anyone interested in greater detail of the theoretical aspects of controlled reproduction as well as broodstock management of both – domesticated and wild percids – is encouraged to see also Fontaine et al. (2015) and Żarski et al. (2015).

Photo-Thermal Manipulations

In the practice of controlled reproduction, a special attention should be paid to the proper and accurate manipulation of photo-thermal conditions. It is well known that constant light conditions (continuous lamination) may inhibit gonadal maturation in Eurasian perch (Migaud et al. 2003). On the other hand, temperature is a modulatory factor in terms of the speed of the maturation process and it may affect gamete quality (Anguis and Cañavate 2005; Targońska et al. 2014). In case of Eurasian perch, a range between 10–16 °C can be considered as the optimal spawning temperature, whereas the photoperiod usually ranges between 12L:12D and 14L:10D (see Żarski et al. 2015). However, in the case of wild and pond-reared fish it may be recommended to keep the fish in constant dimness or lighting (during the illumination period) should be scattered and of low intensity (up to 300 lx on the water surface). Modulation of the temperature during controlled reproduction should be rapid.

Practical Advice
After hormonal injection it is better to keep lower temperatures rather than those too high. Elevated temperatures may negatively affect egg quality (Targońska et al. 2010, 2012).

Theoretical Background of Hormonal Stimulation

The stimulation of the final phases of gamete maturation is possible by acting on the hypothalamic-pituitary-gonadal (HPG) axis with exogenous factors (Mylonas and Zohar 2000). In case of hatchery practice, the hormonal therapy involves methods allowing control over the maturation process by influencing two elements of HPG axis. First is to stimulate the secretion of endogenous gonadotropins from the pituitary by the application of preparations containing gonadotropin releasing hormones (GnRH). The second involves the stimulation of gonadal steroidogenesis by the application of exogenous gonadotropins (GtH) (Donaldson 1996; Mylonas et al. 2010).

In case of Eurasian perch a variety of hormonal preparations were experimentally tested (for details see Żarski et al. 2015). Based on the published data, it can be concluded that two main types of hormonal preparations were found to be the most effective in both, males and females. These include GnRH analogues (GnRHa) and human chorionic gonadotropin (hCG). Therefore, hereafter only the application of these two preparations will be considered.

Important

For hormonal stimulation of Eurasian perch there is no necessity of using the dopamine antagonists as the application of the active substances (GnRH or hCG) alone was found to be suitable enough in induction of ovulation. Since the application of dopamine antagonist was never reported to have any positive effect on the spawning effectiveness of Eurasian perch this aspect was omitted in this manual.

Regulations on the Usage of Spawning Agents in Europe

In the practice of hormonal stimulation of Eurasian perch, only peptide-like spawning agents are used. It should be highlighted that there is no need for the application of any steroid hormones (to achieve good results of spawning), application of which is restricted by specific regulations (Directive 2008/97/EC). This results in the possibility of application of recommended preparations (hCG or GnRHa) the same way as regular pharmacological preparations are allowed (Directive 2001/82/EC). In effect, for controlled reproduction of Eurasian perch, it is allowed to use preparations which are registered for fish, other animals or even humans, under the restriction that the treated fish as well as eggs obtained from treated fish will not be designated for human consumption. This means, that for controlled reproduction of Eurasian perch hCG and/or any kind of preparation containing GnRHa can be used only if it is an officially registered preparation.

Important

Before application of hormonal preparation (or any other drugs) contact your local veterinary authorities in order to verify the possibility and specific terms of usage of particular preparations for a particular purpose.

Where to Inject the Fish?

Fish are usually injected either intramuscularly or intraperitoneally. In the first case, few locations can be chosen such as the dorsal or caudal musculature (Żarski et al. 2015). In this case, the injection is made at the base of the dorsal or anal fin. In case of intraperitoneal injection, hormones are usually administered at the base of the ventral fin which is one of the least scaly parts of the body. There is no proof that the method of injection influences the effectiveness of application of the hormones in any way. However, in case of intramuscular injection any contraction of the musculature may cause a leakage of the administered solution. Therefore, intraperitoneal injection is recommended rather than intramuscular, as is very fast and easy. Upon injection, the needle should be inserted approximately 1–2 cm inside the body (depending on the size of the fish) (Fig. 5.1).

Fig. 5.1 An example of a properly injected needle (the needle was ca. 20 mm long) during intra-peritoneal injection (Photo: S. Krejszeff)

Important
All hormonal injections should be performed with single-use sterile needles and syringes.

Hormonal Stimulation of Females

Preparation of the Hormonal Agent – hCG

Human chorionic gonadotropin is a preparation delivered commercially in the form of an easily soluble substance in a closed sterile vial. Each vial contains a known amount of hCG expressed in IU (international units). Before use, the content of each vial should separately be diluted in a sterile NaCl solution (usually 0.7–0.9 % of NaCl solution). Only after dilution the contents of different vials can be pooled (if necessary).

Practical Advice

Recommended dose of hCG for Eurasian perch is 500 IU per kg of body weight of the fish. Therefore, the dilution of hCG should be performed according to the following formula:

$$X = y / 500$$

Where: X – is the volume of saline solution (in ml) that should be used per one vial; y – is the amount (in IU) of hCG in one vial.

Preparation of the Hormonal Agent – GnRH

Analogues of GnRH are usually delivered in easily soluble crystallized form in sterile vials. Each vial contains a known amount of hormone usually expressed in mg. Before use, the content of each vial should be diluted in a sterile NaCl solution (0.7–0.9 % of NaCl). After dilution the contents of different vials can be pooled, if necessary.

Practical Advice

The recommended dose of GnRH for Eurasian perch is 100 µg per kg of body weight of fish. Therefore, the dilution of GnRH should be performed according to the following formula:

$$X = y / 100$$

Where: X – is the volume of saline solution (in ml) which should be used per one vial; y – is the amount (in µg) of pure GnRHa in one vial.

Hormonal Stimulation of Males

Considering hormonal stimulation of males, the procedure is exactly the same as for females. However, one important aspect is that males should be treated at least 4 days before sperm collection. This is an indispensable condition in order to collect high quality sperm during 'advanced spawning'. The minimum latency time also has a significant influence on sperm volume during the spawning season (Żarski et al. unpublished).

For Eurasian perch, both during and off the spawning season, hormones are recommended to be applied in a single dose:

hCG – 500 IU per kg of body weight,
salmon GnRHa – 50 μg per kg of body weight,
mammalian GnRHa – 100 μg per kg of body weight.

Practical Tips

Photo-Thermal Manipulations

Before injection, it is recommended to keep fish at a temperature that is 2°C lower than the intended spawning temperature. This practice allows to keep the fish at relatively low temperature at which the maturation is slowed down and thus preventing too high desynchronization of the maturation among the females, which can occur in the hatchery (Rónyai and Lengyel 2010; Żarski 2012; Żarski et al. 2015).

After injection, temperature should be increased as fast as possible (fish can safely be transferred, to a different tank/system with a temperature that is 2 °C higher without adverse effects).

It is recommended to keep the same light regime before and after injection.

An Example of the Protocol
1. Until the females reach the maturation stage that allows hormonal treatment, keep them at 10 °C.
2. Immediately after injection, increase the temperature up to 12 °C.
3. Keep the fish in constant dimness and water temperature at 12 °C until the fish spawn (see also Appendix 1)

Hormonal Stimulation

Necessary items:

– Anesthetics and all necessary elements needed for safe anesthesia of the fish (see Chap. 3).
– Hormonal preparations (hCG or GnRHa).
– Sterile saline solutions (0.9 % solution of NaCl).
– 1-ml and 2-ml syringes with needles (needle dimensions: 0.7 mm diameter, 30 mm length).
– Scale with an accuracy of ± 1 g (maximum ±10 g).
– Cloth.
– Manipulation table (benchtop).

The procedure:

1. Dilute the hormonal agent in the saline solution so that the required dose of hormone is contained in 1 ml of the solution.
2. Anesthetize the fish until it reaches surgical anesthesia.

3. Weight the fish.
4. Inject the fish with the hormonal preparation of the required dose.
5. Recover the fish in a recovery tank (see the Chap. 3 on anesthesia procedure).
6. Place the fish into a saline bath for 5 min.
7. Transfer the fish into a holding tank.

Important

In the case of hormonal induction of ovulation in Eurasian perch during the spawning season, a single injection is recommended. For the hormonal stimulation during 'advanced spawning', see Chap. 11.

References

Anguis V, Cañavate JP (2005) Spawning of captive Senegal sole (*Solea senegalensis*) under a naturally fluctuating temperature regime. Aquaculture 243:133–145. doi:10.1016/j. aquaculture.2004.09.026

Baras E, Kestemont P, Mélard C (2003) Effect of stocking density on the dynamics of cannibalism in sibling larvae of *Perca fluviatilis* under controlled conditions. Aquaculture 219:241–255. doi:10.1016/S0044-8486(02)00349-6

Donaldson EM (1996) Manipulation of reproduction in farmed fish. Anim Reprod Sci 42:381–392. doi:10.1016/0378-4320(96)01555-2

Fontaine P, Wang N, Hermelink B (2015) Broodstock management and control of reproductive cycle. In: Kestemont P, Dąbrowski K, Summerfelt RC (eds) Biology and culture of percid fishes, principles and practices. Springer Netherlands, Dordrecht , p 958

Kestemont P, Jourdan S, Houbart M, Mélard C, Paspatis M, Fontaine P, Cuvier A, Kentouri M, Baras E (2003) Size heterogeneity, cannibalism and competition in cultured predatory fish larvae: biotic and abiotic influences. Aquaculture 227:333–356

Mañanós E, Duncan N, Mylonas CC (2008) Reproduction and control of ovulation, spermiation and spawning in cultured fish. In: Cabrita E, Robles V, Herráez MP (eds) Methods in reproductive aquaculture. CRC Press, Boca Raton, pp 3–80

Migaud H, Mandiki R, Gardeur JN, Kestemont P, Bromage N, Fontaine P (2003) Influence of photoperiod regimes on the Eurasian perch gonadogenesis and spawning. Fish Physiol Biochem 28:395–397. doi:10.1023/B:FISH.0000030604.04618.d7

Müller-Belecke A, Zienert S (2008) Out-of-season spawning of pike perch (*Sander lucioperca* L.) without the need for hormonal treatments. Aquac Res 39:1279–1285. doi:10.1111/j.1365-2109.2008.01991.x

Mylonas CC, Zohar Y (2000) Use of GnRHa-delivery systems for the control of reproduction in fish. Rev Fish Biol Fish 10:463–491. doi:10.1023/A:1012279814708

Mylonas CC, Fostier A, Zanuy S (2010) Broodstock management and hormonal manipulations of fish reproduction. Gen Comp Endocrinol 165:516–534. doi:10.1016/j.ygcen.2009.03.007

Rónyai A, Lengyel SA (2010) Effects of hormonal treatments on induced tank spawning of Eurasian perch (*Perca fluviatilis* L.). Aquac Res 41:e345–e347. doi:10.1111/j.1365-2109.2009.02465.x

Targońska K, Kucharczyk D, Kujawa R, Mamcarz A, Żarski D (2010) Controlled reproduction of asp, *Aspius aspius* (L.) using luteinizing hormone releasing hormone (LHRH) analogues with dopamine inhibitors. Aquaculture 306:407–410

Targońska K, Żarski D, Müller T, Krejszeff S, Kozłowski K, Demény F, Urbányi B, Kucharczyk D (2012) Controlled reproduction of the crucian carp *Carassius carassius* (L.) combining temperature and hormonal treatment in spawners. J Appl Ichthyol 28:894–899. doi:10.1111/jai.12073

Targońska K, Żarski D, Kupren K, Palińska-żarska K, Mamcarz A, Kujawa R, Skrzypczak A, Furgała-Selezniow G, Czarkowski TK, Hakuć-Błażowska A, Kucharczyk D (2014) Influence of temperature during four following spawning seasons on the spawning effectiveness of common bream, *Abramis brama* (L.) under natural and controlled conditions. J Therm Biol 39:17–23. doi:10.1016/j.jtherbio.2013.11.005

Żarski D (2012) First evidence of pheromonal stimulation of maturation in Eurasian perch, *Perca fluviatilis* L., females. Turk J Fish Aquat Sci 12:771–776

Żarski D, Horváth A, Held JA, Kucharczyk D (2015) Artificial reproduction of percid fishes. In: Kestemont P, Dąbrowski K, Summerfelt RC (eds) Biology and culture of percid fishes, 1st edn. Springer Netherlands, Dordrecht, pp 123–161

Collection of Gametes

6

Characteristics of Gametes

It is a unique feature of Eurasian perch that eggs are spawned as a cylindrical, sleeve-like structure, which is frequently described in the literature as a strand or ribbon (Treasurer and Holliday 1981; Riehl and Patzner 1998; Probst et al. 2009; Formicki et al. 2009). This structure comprises of a thick, jelly-like cover, directly surrounding each egg (described in detail by Formicki et al. 2009) (Fig. 6.1). Interestingly, the size of the ribbon (width and length) is highly correlated with the size of the female as well as with the number of eggs contained in it (Table 6.1). This feature allows for rough estimation of number of eggs, if the ribbon is collected from the wild, or the tank spawning is performed (as described by Rónyai and Lengyel 2010). The 'dry eggs' (before contact with water) are tightly located within the strand structure (Fig. 6.2). They are entirely transparent and contain a well visible oil droplet. The swollen eggs of Eurasian perch are characterized by a thick jelly-like coat that surrounds the round yolk globule with a well-visible large oil droplet. Between the outer layer and the yolk globule, the perivitelline space allows free rotation of the developing embryo inside the egg.

The spermatozoa of fish are highly simplified cells that are activated following exposure to water (aquasperm) and retain their ability to move and fertilize for a short period of time (approximately 1 min.). The general pattern of most teleost spermatozoa is uniform, although, differences can be detected among taxonomic groups. According to the general pattern these cells have an almost round or slightly elongated head which contains the nucleus, a more or less developed midpiece with one or several mitochondria, that provide energy for movement and a tail region which is called flagellum or axoneme (Stoss 1983). Spermatozoa of the perch have an assymetrically positioned head compared to the axis of the flagellum and a very small midpiece that contains only one mitochondrion (Lahnsteiner et al. 1995).

© The Authors 2017
D. Żarski et al., *Controlled Reproduction of Wild Eurasian Perch*, SpringerBriefs in Environmental Science, DOI 10.1007/978-3-319-49376-3_6

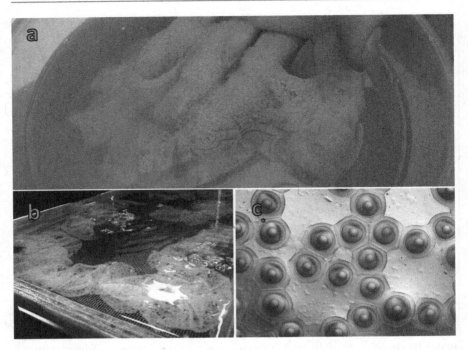

Fig. 6.1 The macro-view of Eurasian perch egg ribbons (**a** and **b**) and microscopic presentation of the egg-ribbon structure with eggs attached to each other (**c**) (Photo: **a** – S.Krejszeff, **b** – D. Żarski, **c** – K. Palińska-Żarska)

Table 6.1 The relation between the length of the ribbon and the number of eggs contained in it

Length of the egg ribbon (cm)	50	75	100	125	150	175	200
Number of eggs (×10³)	6	12	17.5	23.5	29.5	35.5	41.5

According to Gillet et al. (1995)

Determination of the Moment of Ovulation

In the case of Eurasian perch females, recognition of the moment when eggs can be collected is one of the crucial aspects of controlled reproduction. It should be emphasized, that perch females can release eggs spontaneously into the tank, even without the presence of males (Rónyai and Lengyel 2010; Żarski et al. 2011). This is especially the case for wild and pond-reared fish. In the case of cultured fish, sometimes the phenomenon of overripening can be observed (Migaud et al. 2004). This indicates that domesticated fish can sometimes retain mature eggs inside the ovary that. However, the mechanisms behind this phenomenon are still unclear. Nonetheless in both cases, recognition of the moment of ovulation is one of the obstacles in controlled reproduction of this species.

Generally, in Eurasian perch the latency time following hormonal treatment ranges between a few hours to 7 days (Żarski et al. 2015) and is strictly dependent

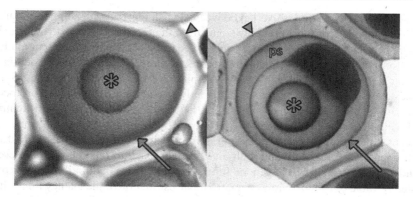

Fig. 6.2 Ovulated Eurasian perch eggs before (*left*) and after (*right*) activation. *Arrows*: emplacement of the internal layer of the chorion, *arrowheads*: outer layer of the *zona radiata externa*, *asterisks*: oil droplets, *ps*: perivitelline space in a swollen egg (for details see Schaerlinger and Żarski 2015) (Photo: D. Żarski)

Table 6.2 Latency time between hormonal treatment and ovulation in Eurasian perch females at 12 °C

Maturation stage at the moment of hormonal injection	Latency time (following hormonal injection)	
	h	days
Stage I	104–168	6–7
Stage II	71–115	5
Stage III	55–76	4
Stage IV	38–56	3
Stage V	24–49	2
Stage VI	6–26	1

on the maturation stage of the fish. That is why, the determination of maturation stage of each female is a very important step of the entire spawning operation. Generally, as described by Żarski et al. (2011), at 12 °C each maturation stage should occur approximately at 24-h intervals. Considering the practical evaluation performed later on, with fish from different populations (Żarski et al. unpublished data), latency time may be predicted according to the following table (Table 6.2):

Important

The latency times described in Table 6.2 refer to a spawning performed at 12 °C. Lower or higher temperatures will cause lengthening or shortening of the latency time, respectively. Also, the latency time may differ between different populations as well as culture conditions.

According to the data presented in Table 6.2 one should keep in mind that the latency time (depending on the type of hormonal preparations used as well as the population) may sometimes overlap between the fish representing neighboring maturational stages. This most probably stems from the fact that oocytes do not usually mature perfectly simultaneously (for details see Chap. 4). There is a high individual variation among particular females in terms of stress sensitivity and reaction to hormonal preparations. Also, it must be emphasized that fish of different sizes and domestication levels may also react differently to particular reproductive protocols (Krejszeff et al. 2009, 2010; Cieśla et al. 2013) which creates additional difficulties in the prediction of the moment of ovulation. Therefore, the latency of a fish at a particular maturational stage following hormonal injection should be verified for each population separately and with a critical approach to this theoretical recommendation. Nevertheless, the general timing of final oocyte maturation process presented in Table 6.2 would allow an approximate estimation of the latency time when a particular population is spawned for the first time.

One of the widest time windows between hormonal injection and ovulation (approx. 64 h) was noted in fish at stage I. This is probably due to the fact that stage I may represent the very beginning of the FOM as well as the final stages of vitellogenesis (Żarski et al. 2011, 2012). It can be hypothesized, that after the hormonal injection the oocytes must sometimes complete vitellogenesis before entering the process of FOM. However, it is only a theoretical assumption which requires more detailed investigation in the future.

Prevention of Spontaneous Ovulation

The phenomenon of spontaneous ovulation (spontaneous release of eggs from the ovary) in percids created the need for development of methods of retaining eggs in the ovary. In case of the Eurasian perch, successful prevention of the egg release is important also from the perspective of manipulation safety, because fish may sometimes release eggs during regular handling procedures (e.g. during netting and removal of fish from the holding tank). So, even though the moment of ovulation can be predicted precisely and/or recognized at the right moment in a particular individual, there is still a risk of losing eggs in the water. Although, there is still a chance to fertilize eggs released spontaneously during manipulations (for details see Chap. 9), from the perspective of controlled fertilization, any method that allows prevention of egg loss is a very useful tool.

In case of Eurasian perch, only one method was reported to be applicable in the hatchery, which is a gentle suture of the genital papilla with a surgical thread. Such a hatchery technique is commonly applied in e.g. common carp, *Cyprinus carpio* L., aquaculture for the same purpose (Woynarovich and Horvath 1980). However, in contrast to common carp in case of Eurasian perch this method does not involve a tight suture of the genital papilla, only a slight closure of the genital region. With this a procedure it is much easier to keep the eggs of Eurasian perch that form a ribbon inside the ovary than in other species that spawn batches of single eggs. A recommended way of suture is presented in Figs. 6.3 and 6.4.

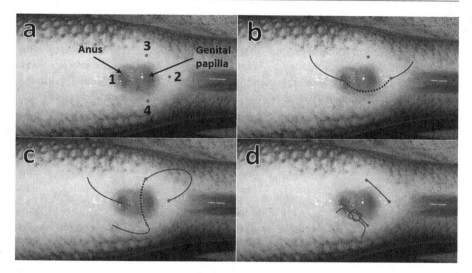

Fig. 6.3 The method of suturing the urogenital pore (here referred to as 'genital papilla') in Eurasian perch female: (**a**) – four *red dots* indicate the spots of puncture with the needle where the numbers next to each dot indicate the order of punctures (numbers *1* and *3* indicate spots where the needle enters the body, numbers *2* and *4* indicate spots where the needle exits); (**b**) –the first suture (the needle enters the anus and exits just next to the beginning of the anal fin of the fish); (**c**) –the second suture (the needle enters and exits both sides of the body, approx. 5 mm from the genital papilla), (**d**) – the thread is tightened and the knot should be made on the ends of the thread at puncture numbers *1* and *4* (Photo: S. Krejszeff)

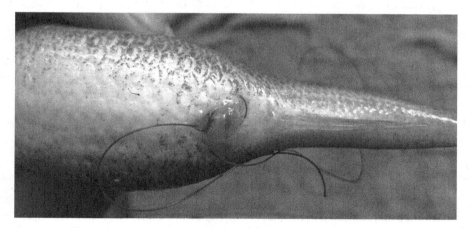

Fig. 6.4 Female of Eurasian perch with sutured (before tightening the thread) genital papilla (Photo: S. Krejszeff)

One of the main bottlenecks for application of the sewing method routinely is its labor intensity, which can possibly generate increment of the production costs. However, the reason for applying this method is the controlled reproduction of valuable females, which are intended to be crossed with particular male and/or males. Also, in the case of the commercial practice, such a solution can be beneficial in the case of spawning the most valuable broodstock or particular females, especially the biggest ones. Therefore, despite the relatively high efforts needed, the sewing method may be a crucial point in specific breeding programs as well as in the case of scientific activities, where eggs from particular females are needed.

Important
The suture of the genital papilla is a harmful manipulation. Before deciding to follow this procedure contact your local veterinary authorities and/or local ethical committee for the necessary permissions and/or terms and conditions of using this method.

Practical Advice
1. Once the genital papilla is sutured and freed (while checking ovulation) it is not recommended to re-suture it again, therefore before the removal of the suture make sure that eggs can be stripped without further complications.
2. Always use clean and possibly sterile thread and needles as well as sharp needles; it is recommended to use surgical thread with a curved needle attached to it which can easily be acquired from pharmacies or medical retail.
3. Adjust needle size to the size of the fish.
4. The thread cannot be too thin as it can cut tissues.

Recognition of the Moment of Ovulation

Regardless of whether the fish were sutured or not, the recognition of the moment when the eggs can be stripped is crucial. Currently, no reliable method exists for its proper recognition, apart from the subjective evaluation by experienced farmers. Generally, the rule is that if eggs are not released easily they should not be stripped. It has to be emphasized that too early stripping can be much more harmful for the fish (much higher pressure on the abdomen is necessary) and the eggs obtained before they are fully ready to be stripped are very often of lower quality. However, in case of Eurasian perch ovulated eggs (those that are ready to be stripped) are very often separated from the outer environment by a specific 'membrane' in the genital papilla (Fig. 6.5a). In many cases the readiness can be over- or underestimated. In the first case the pressure on the abdomen (while the eggs are intended to be stripped)

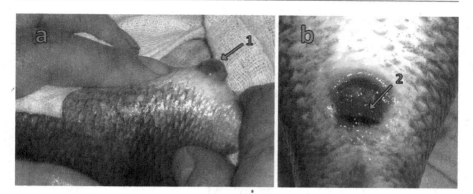

Fig. 6.5 A view of the genital area of a Eurasian perch female: (**a**) – genital membrane (indicated by *arrow 1*); (**b**) – genital opening of the female after breaking of the genital membrane and with eggs that cannot be stripped yet (*arrow 2* indicates the initial part of the eggs, which is in direct contact with the outer environment) (Photo: D. Żarski)

causes breaking of the membrane and the genital opening may come in direct contact with the outer environment which can bring the initial part of the egg-ribbon in contact with water, as well (Fig. 6.5b). However, usually this does not create any problems with egg quality and their fertilizability and only the initial part of the ribbon should be removed (because eggs can undergo activation). In the case when the readiness of the eggs is underestimated (the eggs are not stripped although they should be) there is high risk that the female will release the eggs spontaneously, if not sutured.

> **Practical Advice**
> It is better to store the eggs inside the fish for a few hours more (if the fish is sutured) instead of stripping the eggs too early (for details see Chap. 8).

Collection of Eggs

When the females are ready to spawn, the fish should be dried gently in the area of the genital pore (to avoid contact of the eggs with water) and the ribbon should be stripped directly into a dry container (Fig. 6.6). Just after stripping, the eggs should immediately be covered in order to prevent them from drying. The most convenient is to use small plastic containers (0.5–1 l total volume) with a lid, which are easy to handle, store and can easily be placed in a refrigerator. It is recommended to use one container for each female. Therefore, the proper number of containers should be guaranteed.

From the commercial point of view one of the most important parameter is the fecundity of the females, which allows to predict the number of eggs possible to be obtained from particular female and/or from the entire broodstock available.

Fig. 6.6 Eggs of Eurasian perch during collection. Eggs are of high quality, they are released without any assistance as an example of the proper moment of egg collection (Photo: S. Krejszeff)

However, the already published data indicates, that the fecundity differs considerably between the populations and is not linearly correlated with the size of the fish (Fig. 6.7). Nonetheless, for rough estimation the data presented in Table 6.3 can be used.

Collection of Sperm

In the case of Eurasian perch, there is no problem with sperm collection since males usually release an ample volume of sperm. Therefore, the sperm is usually obtained by regular stripping which causes a flow of the sperm from the sperm duct. The sperm is usually collected into a dry syringe (Fig. 6.8), although in some cases laboratory pipettes are used.

The necessary items:

– Anaesthesia equipment (as described in Chap. 3),
– Syringe (2 ml),
– Cloth,
– Paper towel,

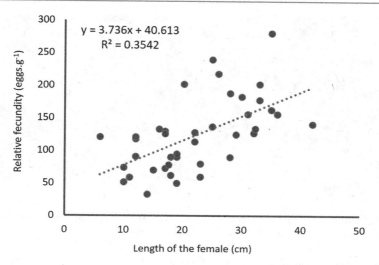

Fig. 6.7 The relationship between the length of the females and relative fecundity (number of eggs per gram of female's body weight) according to different authors (Data taken from Thorpe (1977); Fontaine et al. (2016))

Table 6.3 Reproductive parameters of Eurasian perch females. The data presented are the average values (as well as the limits) calculated on the base of the available literature as well as unpublished data and should be considered only as approximate

	Average	Min	Max
Relative fecundity (number of eggs per g of female's body weight)	123	33	281
Number of eggs per g[a]	500	350	700
Gonadosomatic index (% of body weight)[b]	25	19	31

[a]The value refer to the number of eggs per 1 g of dry eggs, before their contact with water, obtained after stripping
[b]The value refer to the percentage of dry eggs obtained by striping in relation to the wet body weight of the female, before eggs stripping

Practical Advice
– The fish should not be dried completely to avoid wiping off the mucus; gentle wiping of the body with a humid cloth is usually sufficient.
– Mucus does not activate eggs, so do not bother if the mucus comes in contact with the eggs.
– Cover the entire head together with the gill cavity with a cloth as during stripping large volumes of water can be released from it and find their way into the container with eggs.
– Pay attention not to strip eggs in a warm room and into a warm container; the best temperature for manipulation of the stripped eggs is maximum 15 °C.
– Just after the stripping, if eggs are not intended to be fertilized immediately, place the eggs into the storage place (for short-term storage – for details see Chap. 7).

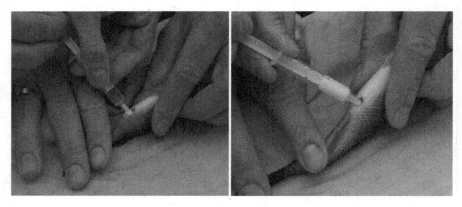

Fig. 6.8 Collection of sperm from Eurasian perch into a dry syringe (beginning and the end of stripping – on the *left* and *right*, respectively) (Photo: Z. Bokor)

– Tubes or other containers for sperm storage.

Procedure of sperm collection:

– Anesthetize the males (follow the procedures described in Chap. 3).
– Put the male on a wet cloth.
– Gently wipe the genital area only with a paper towel.
– Place the syringe to the genital aperture.
– Start to strip the fish gently with an abdominal massage.
– Collect the released sperm into the syringe.
– Put the sperm into a tube or container.
– Immediately after collection put the sperm into a chilled place (e.g. refrigerator or container with melting ice).

Practical Advice
Sperm intended for short-term storage (for more than a few hours) should be collected with a catheter which, by reduction of contamination of sperm with urine, usually improves sperm quality parameters and possibly storage duration (e.g. Sarosiek et al. 2016).

References

Cieśla M, Jończyk R, Gozdowski D, Śliwiński J, Rechulicz J, Andrzejewski W (2013) Changes in ide *Leuciscus idus* (L.) females' reproductive parameters after stimulation with carp pituitary homogenate (CPH) and Ovopel: the effect of domestication? Aquac Int 22:77–88. doi:10.1007/s10499-013-9668-z

Fontaine P, Abdulfatah A, Teletchea F (2016) Reproductive biology and environmental determinism of perch reproductive cycle. In: Couture P, Pyle G (eds) Biology of perch. CRC Press, Boca Raton, pp 167–192

Formicki K, Smaruj I, Szulc J, Winnicki A (2009) Microtubular network of the gelatinous egg envelope within the egg ribbon of European perch, *Perca fluviatilis* L. Acta Ichthyol Piscat 39:147–151. doi:10.3750/AIP2009.39.2.10

Gillet C, Dubois JP, Bonnet S, Lacustre H, Nr IA (1995) Influence of temperature and size of females on the timing of spawning of perch, *Perca fluviatilis*, in Lake Geneva from 1984 to 1993 width (mm) length – width. Environ Biol Fish:355–363

Krejszeff S, Targońska K, Żarski D, Kucharczyk D (2009) Domestication affects spawning of the ide (*Leuciscus idus*)-preliminary study. Aquaculture 295:145–147. doi:10.1016/j. aquaculture.2009.06.032

Krejszeff S, Targońska K, Żarski D, Kucharczyk D (2010) Artificial reproduction of two different spawn-forms of the chub. Reprod Biol 10:67–74

Lahnsteiner F, Berger B, Weismann T, Patzner RA (1995) Fine structure and motility of spermatozoa and composition of the seminal plasma in the perch. J Fish Biol 47:492–508. doi: http:// dx.doi.org/10.1006/jfbi.1995.0154

Migaud H, Fontaine P, Kestemont P, Wang N, Brun-Bellut J (2004) Influence of photoperiod on the onset of gonadogenesis in Eurasian perch *Perca fluviatilis*. Aquaculture 241:561–574. doi:10.1016/j.aquaculture.2004.07.031

Probst WN, Stoll S, Hofmann H, Fischer P, Eckmann R (2009) Spawning site selection by Eurasian perch (*Perca fluviatilis* L.) in relation to temperature and wave exposure. Ecol Freshw Fish 18:1–7. doi:10.1111/j.1600-0633.2008.00327.x

Riehl R, Patzner RA (1998) Minireview: the modes of egg attachment in teleost fishes. Ital J Zool 65:415–420

Rónyai A, Lengyel SA (2010) Effects of hormonal treatments on induced tank spawning of Eurasian perch (*Perca fluviatilis* L.). Aquac Res 41:e345–e347. doi:10.1111/j.1365-2109.2009.02465.x

Sarosiek B, Dryl K, Krejszeff S, Żarski D (2016) Characterization of pikeperch (*Sander lucioperca*) milt collected with a syringe and a catheter. Aquaculture 450:14–16. doi:10.1016/j. aquaculture.2015.06.040

Schaerlinger B, Żarski D (2015) Evaluation and improvements of egg and larval quality in percid fishes. In: Kestemont P, Dabrowski K, Summerfelt RC (eds) Biology and culture of percid fishes. Springer Netherlands, Dordrecht, pp 193–223

Stoss J (1983) Fish gamete preservation and spermatozoan physiology. In: Hoar WS, Randall WH, Donaldson EM (eds) Fish physiology. Academic Press, New York, pp 305–350

Thorpe J (1977) Synopsis of biological data on the perch *Perca fluviatilis* Linnaeus, 1758 and *Perca flavescens* Mitchill, 1814. FAO Fish Synop:1–138

Treasurer JW, Holliday FGT (1981) Some aspects of the reproductive biology of perch *Perca fluviatilis* L. A histological description of the reproductive cycle. J Fish Biol 18:359–376. doi:10.1111/j.1095-8649.1981.tb03778.x

Woynarovich E, Horvath L (1980) The artificial propagation of warm-water finfishes – a manual for extension. Food and Agriculture Organization of the United Nations, Rome

Żarski D, Bokor Z, Kotrik L, Urbanyi B, Horváth A, Targońska K, Krejszeff S, Palińska K, Kucharczyk D (2011) A new classification of a preovulatory oocyte maturation stage suitable for the synchronization of ovulation in controlled reproduction of Eurasian perch *Perca fluviatilis* L. Reprod Biol 11:194–209. doi:10.1016/S1642-431X(12)60066-7

Żarski D, Krejszeff S, Horváth Á, Bokor Z, Palińska K, Szentes K, Łuczyńska J, Targońska K, Kupren K, Urbányi B, Kucharczyk D (2012) Dynamics of composition and morphology in oocytes of Eurasian perch, *Perca fluviatilis* L., during induced spawning. Aquaculture 364–365:103–110. doi:10.1016/j.aquaculture.2012.07.030

Żarski D, Horváth A, Held JA, Kucharczyk D (2015) Artificial reproduction of percid fishes. In: Kestemont P, Dąbrowski K, Summerfelt RC (eds) Biology and culture of percid fishes, 1st edn. Springer Netherlands, Dordrecht, pp 123–161

Short- and Long-Term Storage of Gametes

7

Rationale of Gamete Storage

In many farmed species, fish of opposite sexes usually do not mature perfectly simultaneously. Therefore, it is sometimes necessary to store the gametes for a certain period of time. This makes short- and long-term storage of gametes a very important aspect in aquaculture practice. Even though reproductive protocols typically involve the synchronization of ovulation and spermiation, the possibility of gamete storage may have additional practical applications such as the transfer of gametes between hatcheries. This can become necessary due to problems with reproduction (when the gametes of one sex are of low quality or when the available volumes of sperm or eggs are insufficient) or from the need to crossbreed different populations originating from different locations and the transfer of broodstock is too risky or impossible. On the other hand, long-term storage (i.e. cryopreservation) may allow the preservation of gametes of valuable stocks for several decades.

Short-Term Storage

Both, eggs and sperm undergo irreversible changes after stripping similar to the aging process observed also *in vivo* (i.e. in the gonads of live fish) (Kjørsvik et al. 1990; Bahre Kazemi et al. 2010; Samarin et al. 2011a, b). Interestingly, gametes stored *in vivo* lose their biological function usually much later than the ones stored *in vitro* following manual collection, even if they become mature at the same time. As considering short-term storage, in case of finfish, the storage of sperm is much easier and more feasible than that of eggs. Depending on the species, sperm may easily be stored for a period of time between a few days (approximately 80 h in warm-water species such as the common carp, Ravinder et al. 1997) up to over 1 month (in cold-water species including the rainbow trout, *Oncorhynchus mykiss*, McNiven et al. 1993) following dilution in a specific inactivating solution (Bobe and Labbe 2008; Kowalski et al. 2014) and at low temperatures (about 4 °C). On the

© The Authors 2017
D. Żarski et al., *Controlled Reproduction of Wild Eurasian Perch*, SpringerBriefs in Environmental Science, DOI 10.1007/978-3-319-49376-3_7

contrary, eggs of common carp lose their ability to be fertilized already after 4 h (Lahnsteiner et al. 2001) and eggs of the rainbow trout after about 9 days (Niksirat et al. 2007). However, short-term storage of the eggs applicable in the hatchery is usually much more efficient when eggs are stored without any additions (storage diluents) with the exception of the ovarian fluid, which usually sustains eggs without a negative effect on the storage time (Bobe and Labbe 2008). This suggests, that there are significant interspecies differences in the sensitivity of gametes to storage, whereas the sperm can usually be stored longer. It is important to emphasize, that the duration of storage is also temperature-dependent, and lower temperatures usually allow longer successful storage (Jensen and Alderdice 1984; Bobe and Labbe 2008; Samarin et al. 2011a, b; 2016a). However, according to the published results, the duration of successful storage is generally longer in fish spawning at lower temperatures (such as the rainbow trout) in comparison to species with higher thermal spawning preferences (such as the common carp).

Sperm Cryopreservation

Sperm cryopreservation in fish is an assisted reproduction technology, that involves freezing sperm to −196 °C after dilution with specific extenders and cryoprotective agents (cryoprotectants). Best samples are selected usually according to measurements of motility, concentration and viability. Extenders are mainly prepared from sugars and salts and the osmolality of the solution has an important role in the reversible inactivation of spermatozoa. In freshwater fish, sperm motility is inhibited by hyperosmotic conditions. Cryoprotective agents (mainly alcohols) are intracellular chemicals and their transfer across the cell membrane (a process called equilibration) is a key factor in the success of cryopreservation. Storage of cells in the presence of intracellular cryoprotectants should be limited due to their possible toxic effect. Diluted samples are commonly loaded into straws or cryovials. Freezing can be carried out in insulated boxes filled with liquid nitrogen or in a controlled-rate freezer. Cooling and thawing rates are limiting factors of sperm cryopreservation and are also species specific. While too rapid freezing can lead to damages by the forming ice crystals, slow freezing can result in the dehydration of cells (Cloud and Patton 2008). Post-thaw storage of samples can also be an important factor as it affects the fertilizing capacity of cryopreserved sperm. Currently, there are no known limits of the storage period of cells at ultra-low temperatures. The cryogenic storage of fish sperm can efficiently support controlled reproduction (synchronization of spermiation and ovulation during the spawning season, preservation of high quality samples and simplification of broodstock management). Specific protocols have been developed mainly for freshwater species, particularly salmonids, cyprinids, sturgeons, and catfishes (Cabrita et al. 2010). Anyone interested in the theoretical background of all the aspects of spermatology and sperm cryopreservation of the percid fishes is encouraged to see also Alavi et al. (2015).

In the recent years, several species specific methods of cryopreservation of Eurasian perch sperm were developed from the very basics (extenders, dilution ratio, post-thaw storage, cryopreservation technique, repeatability, fertilization success, etc.) (Rodina et al. 2008; Bernáth et al. 2015a, b), which now were possible to be compiled into one, the most efficient protocol.

Short Term Storage of Eggs of Eurasian Perch

In case of Eurasian perch, short-term storage of eggs is important since the entire procedure of checking of tens of females (including anesthesia, suture of the genital papilla, stripping the eggs and sperm, estimation of eggs quality, verification of the sperm quality, etc.) may last up to 2 h. Additionally, short-term storage of the eggs would facilitate hatchery work when cryopreserved sperm is intended to be used for fertilization or other females are expected to spawn within a few hours. In the latter case, the fertilization procedure can be performed only once allowing a more synchronous incubation of the eggs originating from different females.

Practical Tips for Short-Term Storage of the Eggs of Eurasian Perch
1. Avoid prolonged exposure of stripped eggs to high humidity
2. Just after stripping, cover the eggs with a tight lid
3. Store the eggs from each female in a separate container
4. Follow the recommendations on the duration of egg storage presented in Table 7.1.

Important
The duration of effective short-term storage may be shorter for eggs with fragmented ribbons or with lowered quality. For more details please see also Samarin et al. (2016b)

Short-Term Storage of Sperm

During controlled reproduction of Eurasian perch, sperm very often needs to be stored for a short (or long) period of time. Males can be stripped earlier than females, thus, the quality and fertilizing capacity of milt needs to be preserved preserved

Table 7.1 Maximum duration of short-term *in vitro* storage of Eurasian perch eggs ('dry eggs') following stripping

Temperature of storage	Time of storage (h)
4	18
8	18
12	12

According to Samarin et al. (2016b)

until eggs are available. In perch, this period can last several minutes, hours, or even days. For different purposes, various extenders were developed for Eurasian perch. For a rapid quality assessment (few minutes or some hours) freshly stripped milt can be stored after dilution in a simple buffered salt solution (modified Lahnsteiner's immobilizing solution, Lahnsteiner 2011; Bernáth et al. 2015b). However, if the sperm of Eurasian perch needs to be stored for a longer period (even a few days) the sperm should be diluted in a specific, highly ionic and buffered solution, such as Kobayashi solution (Kobayashi et al. 2004; Sarosiek et al. 2013). For both purposes, extenders are crucial to mitigate the negative effects of contamination of sperm with urine, feces and blood.

Important

In the modified Lahnsteiner's immobilizing solution, sperm can be stored for 6 h in cooled conditions (at 4 °C). Using the Kobayashi solution (Kobayashi et al. 2004; Sarosiek et al. 2013), sperm can be preserved for as much as 48 days. Oxygen supply needs to be provided inside the tubes or containers. For several hours, tubes can stay open, while during storage for several days, oxygen should be provided into closed tubes. During longer storage it is recommended to monitor the dehydration of sperm samples and mix them at intervals of several hours. If the samples are too concentrated, pipetting and dilution of sperm can be facilitated by cropping the ends of pipette tips. It is important to change the tips between two samples. Undiluted sperm can be stored successfully for maximum 1 h (Bernáth et al. 2015b).

Necessary items:

– Scissors (to cut the end of the tip if it is necessary)
– Refrigerator or controlled, cooled conditions (4 °C)
– modified Lahnsteiner's immobilizing solution (150 mM NaCl, 5 mM KCl, 1 mM $MgSO_4 \times 7H_2O$, 1 mM $CaCl_2 \times 2H_2O$, 20 mM Tris, pH 8) (Lahnsteiner 2011)
– Kobayashi solution (130 mM NaCl, 40 mM KCl, 2.5 mM $CaCl_2 \times 2H_2O$, 1.5 mM $MgCl_2 \times 6H_2O$, 2.5 mM $NaHCO_3$, pH 9.5) (Kobayashi et al. 2004)
– Eppendorf tubes, test tubes (with caps), Petri dishes (both sides to have cover)
– Pipettes and tips (in a size according to the amount of sperm and solutions)

Procedure of sperm storage:

1. For undiluted storage, keep the sperm at 4 °C directly after stripping
2. For diluted storage, measure 9 units of extender (type of solution: according to the storage time, see above) into an Eppendorf tube, test tube, Petri dish (according the volume of the sample, Sarosiek et al. 2013)
3. Pipette 1 unit of sperm directly after stripping into 9 units of solutions (dilution ratio 1:9, Sarosiek et al. 2013).
4. Keep the diluted samples at 4 °C

Cryopreservation and Thawing

Cryopreservation allows the storage of sperm collected during 'advanced spawning' until use for hatchery propagation. The best samples (selected by CASA analysis, see in Chap. 8) can be stored in liquid nitrogen without the reduction of quality. An adequate freezing and thawing process was developed to reach a high fertilizing capacity with thawed sperm. Sperm can be cryopreserved in a polystyrene (Styrofoam) box filled with liquid nitrogen (for small volumes of sperm or in field conditions) or with a controlled-rate freezer (for large volumes of stripped milt, Bernáth et al. 2016) (Fig. 7.1). A basic ionic extender is suitable for cryopreservation of Eurasian perch sperm (Bernáth et al. 2015b).

> **Important**
> Sperm cannot be stored before cryopreservation for longer than 1 h. According to the concentration of fresh sperm, pipette tips with cropped ends are recommended to use. Sperm diluted in extender and cryoprotectant can be stored for 1 h before cryopreservation at 4 °C. However, after thawing sperm has a limited storage time of maximum 2 h without a decrease of quality at 4 °C (Bernáth et al. 2015b). When working with liquid nitrogen, the use of specific safety gloves (cryogloves), coat and glasses is obligatory. For thawed sperm, the use of a simple ionic solution is recommended for fertilization and motility assessment (50 mM NaCl, pH 8 Lahnsteiner 2011; Bernáth et al. 2015b).

<u>Necessary items:</u>

- Pipettes and tips (in a range of: 10–100, 100–1000 and 1000–5000 µl)
- Scissors (to cut tips and straws)
- Test tubes or Eppendorf tubes for sperm

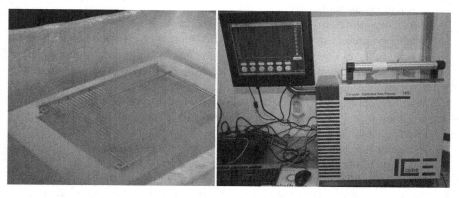

Fig. 7.1 Cryopreservation in a Styrofoam box (on the *left*) and controlled rate freezer (on the *right*) (Photo: G. Bernath)

- Styrofoam box filled with liquid nitrogen for freezing (to a level of 3–4 cm)
- Styrofoam box filled with liquid nitrogen for temporary storage
- Styrofoam frame (height 3 cm)
- Controlled-rate freezer
- Methanol (as cryoprotectant)
- Modified Tanaka extender (137 mM NaCl and 76.2 mM NaHCO3)
- 0.5-mL straws
- Forceps (30 cm long, to manipulate straws in the nitrogen)
- Goblets and canes (to store sperm) in size for 0.5-mL straws
- Storage dewars (store sperm in liquid nitrogen)
- Water bath (for thawing straws)
- Paper towel (dry the straws before placing on the freezing frame)
- Timer (to measure freezing time)
- Markers (to sign straws or goblets or tubes)

Procedure of cryopreservation and thawing:

1. Pipet sperm into Tanaka extender at a ratio 1:10 (use cut-off end tips if it is necessary, change the tips between samples)
2. Add 10 % methanol in final volume into the mixture
3. Mix the diluted sperm gently
4. Label the straws (if it is needed)
5. Adjust a pipette to 500 µL and fix it's tip to the end of the straw (where the polymer is)
6. Push the pipet until the first stop and hold the fixed straw with the pipet in this status
7. Place the open end (without polymer) of the straw into the diluted sample
8. Slowly release the pipette piston and fill the straw in a single long motion
9. Turn the filled, fixed straw and the pipette together upside down. The polymer jellifies upon contact with sperm.
10. Clean the straw with a towel
11. Put on cryogloves for handling nitrogen and frozen straws
12. When cryopreserving in a polystyrene (Styrofoam) box, fill the box with nitrogen to the level of 3–4 cm, and freeze the straws on a frame floating on the surface (3 cm above the level of liquid nitrogen) for 3 min (do not cover the box), (Bernáth et al. 2015b)
13. When cryopreserving with a controlled-rate freezer, set the following cooling program: from 7.5 to −160 °C, cooling rate: 56 °C/min, hold at 160 °C for min, (Bernáth et al. 2015a)
14. After freezing, place the samples into a Styrofoam box filled with nitrogen for 5–10 min (close the box)
15. Label the goblets (if it is necessary) and place them on a cane
16. Place the cane with the goblets into the nitrogen and let the gobles fill up

17. Put the straws into the goblets with a forceps and place the canes into the storage dewars (filled with nitrogen)
18. Before thawing adjust the water bath to 40 °C
19. For thawing remove a straw from the goblet with the forceps
20. With a quick motion, place the straw into the water bath for 13 s
21. Remove the straws from the water bath, wipe them with a paper towel (collect the water with the towel from the open end of the straw)
22. Cover the open end of the straw with your thumb and cut off the end with polimer
23. Release the sperm slowly into a tube
24. Label the tube (if it is necessary)

References

Alavi SMH, Ciereszko A, Hatef A, Křišťan J, Dzyuba B, Boryshpolets S, Rodina M, Cosson J, Linhart O (2015) Sperm morphology, physiology, motility, and cryopreservation in Percidae. In: Kestemont P, Dąbrowski K, Summerfelt RC (eds) Biology and culture of Percid fishes. Springer Netherlands, Dordrecht, pp 163–191

Bahre Kazemi M, Soltani M, Matinfar A, Abtahi B, Pusti I, Mohagheghi Samarin A, Mojazi Amiri B (2010) Biochemical and histological studies of over-ripened oocyte in the Caspian brown trout (Salmo trutta caspius) to determine biomarkers for egg quality. Iran J Fish Sci 9:33–48

Bernáth G, Bokor Z, Kása E, Várkonyi L, Hegyi Á, Kollár T, Urbányi B, Żarski D, Radóczi Ifj J, Horváth Á (2015a) Comparison of two different methods in the cryopreservation of Eurasian perch (Perca fluviatilis) sperm. Cryobiology 70:76–78. doi:10.1016/j.cryobiol.2014.12.003

Bernáth G, Żarski D, Krejszeff S, Palińska-Żarska K, Bokor Z, Król J, Kollár T, Kucharczyk D, Urbányi B, Horváth Á (2015b) Optimization of conditions for the cryopreservation of Eurasian perch (Perca fluviatilis Linnaeus, 1758) sperm. J Appl Ichthyol 31:94–98. doi:10.1111/jai.12740

Bernáth G, Bokor Z, Żarski D, Várkonyi L, Hegyi Á, Staszny Á, Urbányi B, Radóczi Ifj J, Horváth Á (2016) Commercial-scale out-of-season cryopreservation of Eurasian perch (Perca fluviatilis) sperm and its application for fertilization. Anim Reprod Sci 170:170–177. doi:10.1016/j.anireprosci.2016.05.005

Bobe J, Labbe C (2008) Chilled storage of sperm and eggs. In: Cabrita E, Robles V, Herráez P (eds) Methods in reproductive aquaculture: marine and freshwater species. CRC Press, Boca Raton, pp 219–235

Cabrita E, Sarasquete C, Martínez-Páramo S, Robles V, Beirão J, Pérez-Cerezales S, Herráez MP (2010) Cryopreservation of fish sperm: applications and perspectives. J Appl Ichthyol 26:623–635. doi:10.1111/j.1439-0426.2010.01556.x

Cloud J, Patton S (2008) Basic principles of fish spermatozoa cryopreservation. In: Methods in reproductive aquaculture. CRC Press, pp 237–250

Jensen JOT, Alderdice DF (1984) Effect of temperature on short-term storage of eggs and sperm of chum salmon (Oncorhynchus keta). Aquaculture 37:251–265. doi:10.1016/0044-8486(84)90158-3

Kobayashi T, Fushiki S, Ueno K (2004) Improvement of sperm motility of sex-reversed male rainbow trout, Oncorhynchus mykiss, by incubation in high-pH artificial seminal plasma. Environ Biol Fishes 69:419–425. doi:10.1023/B:EBFI.0000022904.35065.e8

Kjørsvik E, Mangor-Jensen A, Holmefjord I (1990) Egg quality in fishes. Adv Mar Biol 26:71–113

Kowalski RK, Cejko BI, Irnazarow I, Szczepkowski M, Dobosz S, Glogowski J (2014) Short-term storage of diluted fish sperm in air versus oxygen. Turk J Fish Aquat Sci 14:831–834. doi:10.4194/1303-2712-v14_3_26

Lahnsteiner F (2011) Spermatozoa of the teleost fish Perca fluviatilis (perch) have the ability to swim for more than two hours in saline solutions. Aquaculture 314:221–224. doi:10.1016/j.aquaculture.2011.02.024

Lahnsteiner F, Urbanyi B, Horvath A, Weismann T (2001) Bio-markers for egg quality determination in cyprinid fish. Aquaculture 195:331–352. doi:10.1016/S0044-8486(00)00550-0

McNiven MA, Gallant RK, Richardson GF (1993) Fresh storage of rainbow trout (Oncorhynchus mykiss) semen using a non-aqueous medium. Aquaculture 109:71–82. doi:10.1016/0044-8486(93)90487-J

Niksirat H, Sarvi K, Mojazi Amiri B, Hatef A (2007) Effects of storage duration and storage media on initial and post-eyeing mortality of stored ova of rainbow trout Oncorhynchus mykiss. Aquaculture 262:528–531. doi:10.1016/j.aquaculture.2006.10.031

Ravinder K, Nasaruddin K, Majumdar KC, Shivaji S (1997) Computerized analysis of motility, motility patterns and motility parameters of spermatozoa of carp following short-term storage of semen. J Fish Biol 50:1309–1328. doi:10.1111/j.1095-8649.1997.tb01655.x

Rodina M, Policar T, Linhart O, Rougeot C (2008) Sperm motility and fertilizing ability of frozen spermatozoa of males (XY) and neomales (XX) of perch (Perca fluviatilis). J Appl Ichthyol 24:438–442. doi:10.1111/j.1439-0426.2008.01137.x

Samarin AM, Amiri BM, Soltani M, Mohammad R (2011a) Effects of storage duration and storage temperature on viability of stored ova of kutum (Rutilus frisii kutum) in ovarian fluid. Afr J Biotechnol 10:12309–12314. doi:10.5897/AJB11.919

Samarin AM, Amiri BM, Soltani M, Nazari RM, Kamali A, Naghavi MR (2011b) Effects of post-ovulatory oocyte ageing and temperature on egg quality in kutum Rutilus frisii kutum. World Appl Sci J 15:14–18

Samarin AM, Blecha M, Uzhytchak M, Bytyutskyy D, Zarski D, Flajshans M, Policar T (2016a) Post-ovulatory and post-stripping oocyte ageing in northern pike, Esox lucius (Linnaeus, 1758), and its effect on egg viability rates and the occurrence of larval malformations and ploidy anomalies. Aquaculture 450:431–438. doi:10.1016/j.aquaculture.2015.08.017

Samarin AM, Żarski D, Palińska-Żarska K, Krejszeff S, Blecha M, Kucharczyk D, Policar T (2016b) In vitro storage of unfertilized eggs of the Eurasian perch and its effect on egg viability rates and the occurrence of larval malformations. Animal:1–6. doi:10.1017/S1751731116001361

Sarosiek B, Cejko BI, Kucharczyk D, Zarski D, Judycka S, Kowalski RK (2013) Short-term storage of perch (Perca fluviatilis L.) milt under cooling conditions. Reprod Biol. doi:10.1016/j.repbio.2012.11.036

Evaluation of Gamete Quality

8

How to Define Gamete Quality?

Gamete quality by definition is the ability of sperm to fertilize and that of eggs to be fertilized which consequently leads to the development of a 'normal embryo' (Bobe and Labbé 2010). In intensive aquaculture, gamete quality is one of the most important aspects directly determining production effectiveness. High-quality gametes ensure the production of high-quality larvae. In case of percid aquaculture, gamete quality is usually very variable and hardly predictable (Żarski et al. 2011, 2012). Therefore attempts had been made to determine the factors responsible for gamete quality. Feeding regimes (Henrotte et al. 2010a, b), environmental conditions (Migaud et al. 2004; Fontaine et al. 2006; Abdulfatah et al. 2011), as well as reproductive protocols (Kucharczyk et al. 1996, 1998; Kouril et al. 1997; Targońska et al. 2014) were listed as the most important factors that have a direct effect on gonadal development, maturation and consequently on gamete quality. This makes the regulation of gamete quality, especially eggs, multifactorial issue.

In the Eurasian perch, the highest variability of spawning effectiveness results usually from the impaired quality of eggs. Therefore, several studies concentrated on the effect of different factors responsible for the developmental competence of eggs, where nutrition and environmental manipulation were the main concerns. However, many other elements can also alter gamete quality such as stress and stress-related factors that influence hormonal economy responsible for the proper course of gonadal maturation (Schreck et al. 2001; Barton 2002; Milla et al. 2015).

In hatchery practice, methods allowing the earliest possible recognition of gamete quality are very important tools. It allows the selection of gametes characterized by the highest quality for further steps of controlled reproduction to ensure high fertilization rate. Low fertilization efficiency leads to many problems as non-developing embryos are sources of infection (mainly bacterial, protozoan and fungal) for otherwise viable embryos during incubation, thus, they can cause a significant reduction in the incubation. This is especially important in the case of Eurasian perch whose eggs are located within the same structure of the ribbon which prevents the

© The Authors 2017
D. Żarski et al., *Controlled Reproduction of Wild Eurasian Perch*, SpringerBriefs in Environmental Science, DOI 10.1007/978-3-319-49376-3_8

separation of viable and non-viable eggs from each other. However, the methods of distinguishing gametes according to quality are very limited.

Egg quality itself as well as factors determining the quality of eggs in percids was recently reviewed by Schaerlinger and Żarski (2015), where anyone interested can find a more detailed theoretical background related with this aspect.

Methods of Evaluation of Gamete Quality

Up to now, the only reliable method allowing impartial evaluation of gamete quality was developed for sperm. The spermatozoa of fish remain inactive until they come into direct contact with water. Therefore, sperm stripped into a dry container (syringe or test tube) can be stored for a certain period of time (for details see the Chap. 7) without losing its fertilizing capacity. However, just after contact with water (or any other activating medium) the cells undergo activation and become motile. Motility parameters (such as motility rate or velocity) and motility duration strongly correlate with sperm quality (Fauvel et al. 2010). This feature is also used in the hatchery practice where subjective evaluation of motility became a standard method of selection of sperm for commercial fertilization purposes (Żarski et al. 2011). The evaluation of egg quality is more difficult and there are very few indicators that allow reliable verification of egg quality (for details see Schaerlinger and Żarski 2015). Although some progress can be observed, still, the most credible method for the evaluation of egg quality is the determination of embryonic survival at a particular developmental stage. This stems from the fact that poor egg quality can result from an improper composition of the yolk (including the fatty acid profile), which is a nutritional source for the embryos and larvae as well as from the molecular status of the egg (maternal mRNA and protein levels) (Castets et al. 2012; Schaerlinger and Żarski 2015). Therefore, investigation of developmental competence in the embryos, as a clear evidence of high egg quality, is the most credible method to be applied in the hatchery practice. However, recently Żarski et al. (2011) reported that some morphological features of eggs, when evaluated prior to fertilization could be a helpful tool in the case of controlled reproduction of Eurasian perch. However, it must be emphasized, that the morphological characteristics of ovulated eggs with high probability will allow to distinguish eggs of the lowest quality, only. There is still no reliable method in the Eurasian perch that would allow to recognize eggs of low quality when morphological characteristics are not altered.

Hatchery-Related Factors Affecting Gamete Quality

Practically all the steps of the controlled reproduction should be considered as potential factors negatively affecting gamete quality, since all reproductive procedures are stressful for fish. In addition, it was proven many times that the type and dose of the hormonal preparations applied also directly influenced egg and sperm quality, including that of the Eurasian perch (Kucharczyk et al. 1996, 1998; Kouril et al. 1997; Targońska et al. 2014). A special attention should be paid to the time of gamete

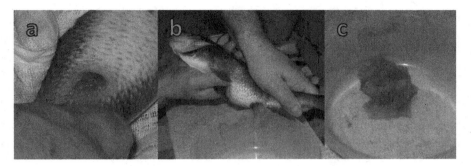

Fig. 8.1 An example of premature egg stripping: (**a**) genital membrane was broken upon control of ripeness (as also shown on Fig. 6.5), (**b**) eggs could be stripped only after high pressure on the abdomen, (**c**) eggs stripped do not exhibit the typical 'ribbon-like' shape and are clumped together (Photo: D. Żarski)

collection. It is a well-known fact that a delayed collection of gametes (following hormonal stimulation) is the reason of obtaining lower quality gametes. This stems from overripening (in case of females, Lahnsteiner 2000; Samarin et al. 2008, 2011) or sperm aging (in case of males, Cejko et al. 2012). On the contrary, in case of percids, there are justified assumptions that gametes collected too early (both – eggs and sperm) are also characterized by lower quality (Fig. 8.1). This probably results from premature gamete collection (when they can already be collected manually but not ready to be fertilized), however, more work is needed to verify these hypotheses. Nevertheless, it is important to highlight that the proper moment of gamete collection is a crucial element of successful reproduction. In effect, from the practical point of view, these two aspects of controlled reproduction should be considered as the ones which may alter spawning effectiveness by influencing gamete quality.

Evaluation of the Egg Quality

Methods and possibilities allowing determination of the biological quality of eggs before fertilization are severely limited. In the case of Eurasian perch, one of the possibilities is to recognize low egg quality by evaluation of morphological features under a stereoscopic microscope. However, there is no method that would allow to indicate high quality of the eggs with good probability.

Practical Advice
For morphological evaluation of egg quality, the magnification of regular light microscopes is usually too high – therefore a stereoscopic microscope is usually necessary. For the commercial purposes, the regular stereo microscope with the ability of the light transmission through the object observed and possibility of adjustment of magnification, is highly advanced. Nowadays, modern stereo microscope (with all the features mentioned) can be possible to purchase for less than 500 EUR (~550 USD).

Morphological Evaluation Off Egg Quality

Fragmentation of Oil Droplets

Fragmentation of oil droplets in ovulated oocytes were found to be a good indicator of low egg quality (Żarski et al. 2011). The evaluation must be performed before contact of eggs with water, or immediately afterwards, because during the swelling process the fragmented oil droplets can merge into a single droplet. However, in many cases fragmented oil droplets can be also found throughout the embryonic development (Fig. 8.2).

> **Important**
> The presence of a single oil droplet is not an indicator of high egg quality.

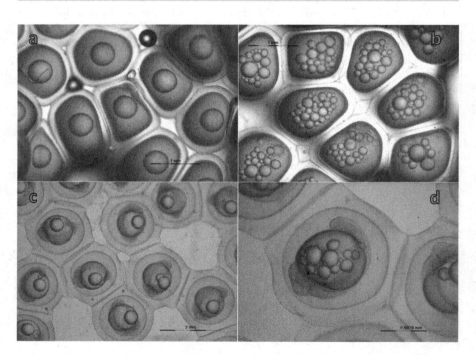

Fig. 8.2   The dry eggs (**a** and **b**) and developing embryos (**c** and **d**) of Eurasian perch: (**a**) dry eggs of high quality with only one oil droplet, (**b**) dry eggs of low quality with fragmented oil droplets, (**c**) developing embryos of high quality with mostly one oil droplet, (**d**) low quality embryo with fragmented oil droplet (According to Żarski et al. 2011) (Photo: D. Żarski)

Morphological Alterations of Various Types

Generally, visual observation (such as color) is too subjective to be considered a reliable quality indicator. It was reported that the white (opaque) eggs within the egg batch or fragmentation of egg-ribbon structure can be considered as an indicator of lowered egg quality (Dabrowski et al. 1994; Castets et al. 2012; Schaerlinger and Żarski 2015). However, despite easy to recognize signs of possibly lowered quality, it is recommended to verify the macroscopic observations under the stereoscopic microscope any possible morphological alterations, in order to confirm the low egg quality and/or verify the percentage of the eggs exhibiting abnormal features. Such a morphological evaluation is usually quite obvious low egg quality can be recognized with a high certainty (Fig. 8.3).

First Cleavages and Blastomere Morphology

Generally, eggs with developmental competence always exhibit their ability to develop until a particular developmental stage (Alix et al. 2013; Schaerlinger and Żarski 2015) that can be noticed during the first cell cleavages of inseminated eggs. Usually it is questionable whether an embryo will finally develop into a 'normal larva' since occurrence of the first division(s) gives information only on whether the eggs are fertilizable or not. However, it was already reported for some finfish species, that blastomere morphology (Shields et al. 1997) can be considered a reliable quality indicator of eggs. Although this aspect has not been studied in case of the Eurasian perch, yet, it can be considered as a possibly applicable method of evaluation of egg quality. However, it must be emphasized, that this aspect still requires further verification of reliability, especially that in some cases abnormal cell cleavages may have no effect on the final success of incubation (see Vallin and Nissling 1998).

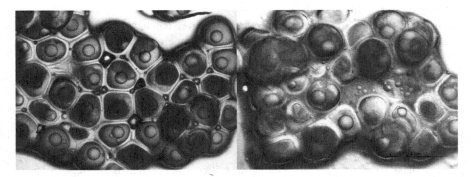

Fig. 8.3 Examples of various morphological alterations in freshly ovulated eggs of low quality, before their contact with water. The lack of transparency is evident in some of the eggs as well as internal damages to yolk structures (compare also with Fig. 8.2a) (Photo: D. Żarski)

Although evaluation of the first divisions or blastomere morphology cannot be performed before fertilization, it can be still a useful tool in hatchery practice and scientific work, especially, that the eggs in certain conditions may be stored for several hours without a negative effect on their quality (see Chap. 7). In order to use such indicators for a quick estimation of egg quality, the following procedure should be applied

Practical Tips on Evaluation of Egg Quality

Necessary items:

− Container with the eggs ready for short-term storage (as described in Chap. 7)
− Stereoscopic microscope
− Petri-dishes
− Activating solution (see Chap. 9)
− Hatchery water
− Good quality (at least 80 % motility rate) sperm sample

Procedure of evaluation:

1. Take an egg sample (from approximately the middle of the egg ribbon) with at least 100 eggs; reserve the remaining eggs for short-term chilled storage (see Chap. 7)
2. Place the egg sample on the Petri-dish
3. Pour activating solution on the Petri-dish with the egg sample
4. Spread the egg sample in the activating solution throughout the Petri-dish
5. 15 s following egg activation add a sperm sample (sperm should be given in excess − increase the dose 10-fold in comparison to the recommended one presented in Chap. 9)
6. Stir the eggs (with e.g. a needle, paying attention not to damage the eggs) for about 3 min in the Petri dish
7. Replace the mixture of water and sperm with pure hatchery water
8. Put the Petri dish in a place where incubation temperature will range between 15 and 16 °C (the highest among the recommended ones will accelerate the commencement of the first cleavages)
9. After approximately 2 h verify the occurrence of first cell cleavages and/or start the verification of blastomere morphology (Fig. 8.4).

Practical Advice
The blastomere morphology should be verified no later than at the stage of 16 cells (after fourth cleavage).

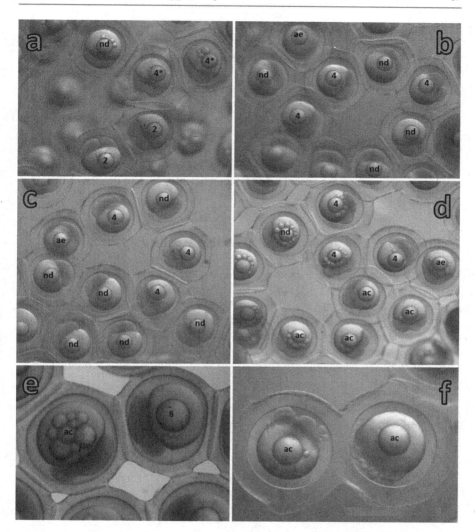

Fig. 8.4 An examples of various cleavage patterns: (**a**) an egg batch in which embryos exhibit an asynchronous pattern of cleavages; (**b, c**) egg batches were all developing embryos exhibit the stage after the second cleavage (4-cell stage) and non-developing as well as altered eggs can be observed; (**d**) an egg batch where mostly abnormal cleavages can be observed, (**e**) egg batch at 8-cell stage, where the one with fragmented oil droplet exhibits abnormal blastomere morphology, (**f**) eggs with abnormal cleavages form a batch of eggs which should exhibit 16-cell stage. Numbers represent the number of blastomeres in a particular egg (*asterisk* indicates beginning of a particular stage), *nd* non-developing eggs, *ae* altered egg (morphological alteration), *ac* abnormal blastomere morphology (Photo: D. Żarski)

Embryonic Survival Rate

From the practical point of view, it is important not only to choose the batch of eggs which will be subjected to fertilization and incubation, but also the verification of the all reproductive procedures. For that purpose fertilization rate determined at the latest possible stages of embryonic development (which is the most reliable egg quality indicator), is typically used. The most reliable stage of verification of egg quality is the eyed eggs stage, when all non-developing eggs are usually non-transparent. Of course sometimes it is necessary to estimate the fertilization rate earlier, but in these cases special attention should be paid to distinguishing live and dead embryos, since eggs can remain transparent until the late phases of development (Fig. 8.5).

Evaluation of the Sperm Quality by the Motility Assessment

During the process, stripped sperm quality is evaluated using a CASA (Computer-assisted Sperm Analysis, Figs. 8.6 and 8.7) system. Different parameters of movement can be recorded and stored with a CASA. Movement of cells can also be estimated visually with a regular light microscope. It is necessary to sort samples according the percentage of motile and immotile cells, thus, the success of fertilization can be improved with fresh and thawed sperm as well. In the cryopreserved sperm of Eurasian perch, an estimated 30–50 % motility decrease is predictable in

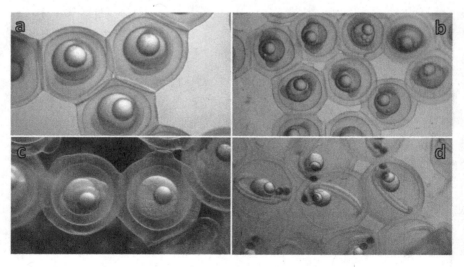

Fig. 8.5 Examples of eggs of Eurasian perch at different phases of embryonic development: (**a**) second-third day of incubation, (**b**) third-fourth day of incubation, (**c**) third-fourth day of incubation where developing (*left*) and non-developing (*right*) embryos are visible, (**d**) embryos at eyed-egg stage (Photo: D. Żarski)

Fig. 8.6 Sperm analysis with a CASA system. The microscope is connected to a computer which uses the specific CASA software (Photo: G. Bernath)

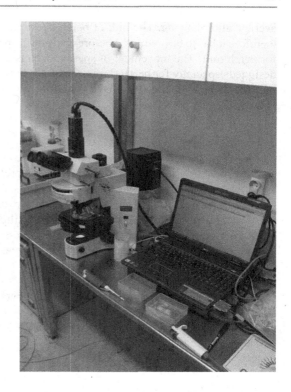

Fig. 8.7 Motile (cells with a track), locally motile (*blue dots*) and immotile (*red dots*) sperm cells recorded with a CASA system (Photo: G. Bernath)

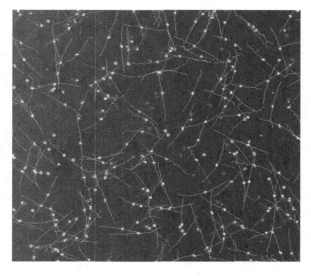

average after thawing. The milt of perch is extremely concentrated (see Table 9.1) therefore it has to be prediluted before analysis in an immobilizing solution designed for perch milt. The movement of spermatozoa is activated with a specific solution designed especially for perch (Lahnsteiner 2011; Bernáth et al. 2015b). After thawing, the sample is already diluted therefore no further dilition is needed before measurement.

Important

Freshly stripped sperm is recommended to be stored on 4 °C to avoid a possible reduction in quality. It is advisable to prepare fresh immobilizing and activating solutions before the measurements, however the diluents can be stored in a freezer. It is important to avoid the drifting of cells during CASA measurements as they can result in false motility readings. Perch sperm can be very concentrated in which case it is necessary to use tips with cropped ends. It is important to change pipette tips between two measurements. A suitable number of cells in the observation field is 200–500. Measurements are advisable to conduct at least in duplicates or triplicates in all samples. After every measurement or activation, the chambers and slides have to be very accurately cleaned with precision wipes (to avoid scratching the surface of the chamber or slide). Sperm cells are able to move for 30 s in average after activation (Bernáth et al. 2015a).

The necessary items:

- Eppendorf tubes for the solutions.
- Markers (to sign tubes).
- Pipettes and tips (1–10 μl and 10–100 μl).
- Modified Lahnsteiner's immobilizing solution (150 mM NaCl, 5 mM KCl, 1 mM MgSO$_4$ × 7H$_2$O, 1 mM CaCl$_2$ × 2H$_2$O, 20 mM Tris, pH 8, (Lahnsteiner 2011; Bernáth et al. 2015a).
- modified Lahnsteiners activating solution (75 mM NaCl, 2 mM KCl, 1 mM MgSO$_4$ × 7H$_2$O, 1 mM CaCl$_2$ × 2H$_2$O, 20 mM Tris, pH 8, (Lahnsteiner 2011; Bernáth et al. 2015a) in a mixture with 0.01 g/mL BSA (Bovine Serum Albumin).
- CASA system or light microscope.
- Counting chambers (for CASA), or regular slides (for light microscopes).
- Styrofoam box with crushed melting ice (sperm storage).
- Precision wipes.
- Scissor (to crop the ends of tips if necessary).

Procedure of analysis:

1. Place the stripped fresh or thawed sperm on the ice
2. Lunch the CASA system or the microscope

3. Place the counting chamber or other slides onto the microscope stage
4. For fresh sperm: pipette 2 µl pure sperm into 100 µl immobilizing solution (sometimes with a tip with a cropped end)
5. Place 20 µl activating solution into the chamber
6. Pipet and mix accurately 1 µl of fresh diluted or thawed sperm into the immobilizing solution
7. Analyze the sample using the CASA system or a regular light microscope within 10 s

References

Abdulfatah A, Fontaine P, Kestemont P, Gardeur J-N, Marie M (2011) Effects of photothermal kinetics and amplitude of photoperiod decrease on the induction of the reproduction cycle in female Eurasian perch Perca fluviatilis. Aquaculture 322–323:169–176. doi:10.1016/j.aquaculture.2011.09.002

Alix M, Schaerlinger B, Ledoré Y, Chardard D, Fontaine P (2013) Developmental staging and deformities characterization of the Eurasian perch, Perca fluviatilis. Commun Agric Appl Biol Sci 78:12–14

Barton BA (2002) Stress in fishes: a diversity of responses with particular reference to changes in circulating corticosteroids. Integr Comp Biol 42:517–525. doi:10.1093/icb/42.3.517

Bernáth G, Bokor Z, Kása E, Várkonyi L, Hegyi Á, Kollár T, Urbányi B, Żarski D, Radóczi Ifj J, Horváth Á (2015a) Comparison of two different methods in the cryopreservation of Eurasian perch (Perca fluviatilis) sperm. Cryobiology 70:76–78. doi:10.1016/j.cryobiol.2014.12.003

Bernáth G, Żarski D, Krejszeff S, Palińska-Żarska K, Bokor Z, Król J, Kollár T, Kucharczyk D, Urbányi B, Horváth Á (2015b) Optimization of conditions for the cryopreservation of Eurasian perch (Perca fluviatilis Linnaeus, 1758) sperm. J Appl Ichthyol 31:94–98. doi:10.1111/jai.12740

Bobe J, Labbé C (2010) Egg and sperm quality in fish. Gen Comp Endocrinol 165:535–548. doi:10.1016/j.ygcen.2009.02.011

Castets M-DD, Schaerlinger B, Silvestre F, Gardeur J-NN, Dieu M, Corbier C, Kestemont P, Fontaine P (2012) Combined analysis of Perca fluviatilis reproductive performance and oocyte proteomic profile. Theriogenology 78(432–442):442–413. doi:10.1016/j.theriogenology.2012.02.023

Cejko BI, Kowalski RK, Żarski D, Dryl K, Targońska K, Chwaluczyk R, Kucharczyk D, Glogowski J (2012) The influence of the length of time after hormonal treatment with [(d-Ala6, Pro9 NEt)-mGnRH+metoclopramide] i.e. Ovopel on barbel Barbus barbus (L.) milt quality and quantity indicators. J Appl Ichthyol 28:249–253. doi:10.1111/j.1439-0426.2011.01923.x

Dabrowski K, Ciereszko A, Ramseyer L, Culver D, Kestemont P (1994) Effects of hormonal treatment on induced spermiation and ovulation in the yellow perch (Perca flavescens). Aquaculture 120:171–180. doi:10.1016/0044-8486(94)90231-3

Fauvel C, Suquet M, Cosson J (2010) Evaluation of fish sperm quality. J Appl Ichthyol 26:636–643. doi:10.1111/j.1439-0426.2010.01529.x

Fontaine P, Pereira C, Wang N, Marie M (2006) Influence of pre-inductive photoperiod variations on Eurasian perch Perca fluviatilis broodstock response to an inductive photothermal program. Aquaculture 255:410–416. doi:10.1016/j.aquaculture.2005.12.025

Henrotte E, Kaspar V, Rodina M, Psenicka M, Linhart O, Kestemont P (2010a) Dietary n-3/n-6 ratio affects the biochemical composition of Eurasian perch (Perca fluviatilis) semen but not indicators of sperm quality. Aquac Res 41:e31–e38. doi:10.1111/j.1365-2109.2009.02452.x

Henrotte E, Mandiki RSNM, Prudencio AT, Vandecan M, Mélard C, Kestemont P (2010b) Egg and larval quality, and egg fatty acid composition of Eurasian perch breeders (Perca fluviatilis) fed different dietary DHA/EPA/AA ratios. Aquac Res 41:e53–e61. doi:10.1111/ j.1365-2109.2009.02455.x

Kouril J, Linhart O, Relot P (1997) Induced spawning of perch by means of a GnRH analogue. Aquac Int 5:375–377

Kucharczyk D, Kujawa R, Mamcarz A, Skrzypczak A, Wyszomirska E (1996) Induced spawning in perch, Perca fluviatilis L. using carp pituitary extract and HCG. Aquac Res 27:847–852. doi:10.1046/j.1365-2109.1996.t01-1-00802.x

Kucharczyk D, Kujawa R, Mamcarz A, Skrzypczak A, Wyszomirska E (1998) Induced spawning in perch, Perca fluviatilis L., using FSH + LH with pimozide or metoclopramide. Aquac Res 29:131–136. doi:10.1046/j.1365-2109.1998.00949.x

Lahnsteiner F (2000) Morphological, physiological and biochemical parameters characterizing the over-ripening of rainbow trout eggs. Fish Physiol Biochem 23:107–118. doi:10.1023/ A:1007839023540

Lahnsteiner F (2011) Spermatozoa of the teleost fish Perca fluviatilis (perch) have the ability to swim for more than two hours in saline solutions. Aquaculture 314:221–224. doi:10.1016/j. aquaculture.2011.02.024

Migaud H, Fontaine P, Kestemont P, Wang N, Brun-Bellut J (2004) Influence of photoperiod on the onset of gonadogenesis in Eurasian perch Perca fluviatilis. Aquaculture 241:561–574. doi:10.1016/j.aquaculture.2004.07.031

Milla S, Douxfils J, Mandiki SNM, Saroglia M (2015) Corticosteroids and the stress response in percid fish. In: Kestemont P, Dąbrowski K, Summerfelt RC (eds) Biology and culture of percid fishes, principles and practices. Springer, Dordrecht, p 891

Samarin AM, Ahmadi MR, Azuma T, Rafiee GR, Amiri BM, Naghavi MR (2008) Influence of the time to egg stripping on eyeing and hatching rates in rainbow trout Oncorhynchus mykiss under cold temperatures. Aquaculture 278:195–198. doi:10.1016/j.aquaculture.2008.03.034

Samarin AM, Amiri BM, Soltani M, Nazari RM, Kamali A, Naghavi MR (2011) Effects of post-ovulatory oocyte ageing and temperature on egg quality in kutum rutilus frisii kutum. World Appl Sci J 15:14–18

Schaerlinger B, Żarski D (2015) Evaluation and improvements of egg and larval quality in percid fishes. In: Kestemont P, Dabrowski K, Summerfelt RC (eds) Biology and culture of percid fishes. Springer, Dordrecht, pp 193–223

Schreck CB, Contreras-Sanchez W, Fitzpatrick MS (2001) Effects of stress on fish reproduction, gamete quality, and progeny. Aquaculture 197:3–24

Shields RJ, Brown NP, Bromage NR (1997) Blastomere morphology as a predictive measure of fish egg viability. Aquaculture 155:1–12. doi:10.1016/S0044-8486(97)00105-1

Targońska K, Szczerbowski A, Żarski D, Łuczyński MJ, Szkudlarek M, Gomułka P, Kucharczyk D (2014) Comparison of different spawning agents in artificial out-of-season spawning of Eurasian perch, Perca fluviatilis L. Aquac Res 45:765–767. doi:10.1111/are.12010

Vallin L, Nissling A (1998) Cell morphology as an indicator of viability of cod eggs – results from an experimental study. Fish Res 38:247–255. doi:10.1016/S0165-7836(98)00157-X

Żarski D, Palińska K, Targońska K, Bokor Z, Kotrik L, Krejszeff S, Kupren K, Horváth Á, Urbányi B, Kucharczyk D (2011) Oocyte quality indicators in Eurasian perch, Perca fluviatilis L., during reproduction under controlled conditions. Aquaculture 313:84–91. doi:10.1016/j. aquaculture.2011.01.032

Żarski D, Krejszeff S, Palińska K, Targońska K, Kupren K, Fontaine P, Kestemont P, Kucharczyk D (2012) Cortical reaction as an egg quality indicator in artificial reproduction of pikeperch, Sander lucioperca. Reprod Fertil Dev 24:843. doi:10.1071/RD11264

In Vitro Fertilization

9

Theoretical Background

In vitro fertilization is a technique involving methods of exposition of mature gametes to a medium which activates both, sperm and eggs (Żarski et al. 2014). In the literature, this medium is called the activating solution (AS) (Krise et al. 1995). At this step of controlled reproduction, human intervention has a main influence on the effectiveness of the process. By adjusting a proper sperm-to-egg ratio, choosing the right AS and performing the exposition of the gametes to each other in a proper way all eggs that have a developmental competence will be fertilized with the highest probability. The *in vitro* fertilization procedure has been applied in finfishes for many years (Radziwoński 1852; Billard et al. 1974; Woynarovich and Woynarovich 1980). However, biological properties of the gametes vary among various species, so the development of the fertilization procedure should be verified for every species separately in order to ensure the highest efficiency (Żarski et al. 2012a; 2014; 2015). This is especially important in the case of new species in aquaculture as well as the species characterized by specific features of the gametes. Both conditions are met by the Eurasian perch which is an emerging aquaculture species that has unique egg structure (for details see Chap. 6).

Technique of *In Vitro* Fertilization

In aquaculture practice, the procedure of *in vitro* fertilization involves mixing of the dry gametes with each other and later activation of the gametes by adding AS (Żarski et al. 2015). However, in the case of Eurasian perch it was found that application of such a procedure (when hatchery water is used as an AS) reduces fertilization rate as compared to the eggs which were first activated with an AS and inseminated 15 s following eggs activation (Żarski et al. 2012a). This probably stems from the fact that the eggs of Eurasian perch, situated within a gelatinous-like cylindro-conical 'ribbon', are very difficult to mix with sperm and additionally eggs

© The Authors 2017
D. Żarski et al., *Controlled Reproduction of Wild Eurasian Perch*, SpringerBriefs in Environmental Science, DOI 10.1007/978-3-319-49376-3_9

usually cover one another. In this case, upon addition of AS the eggs still cover each other and by the time when the eggs are fully dispersed within the AS and the entire ribbon will acquire the shape allowing exposition of all eggs to the AS, the sperm motility is significantly reduced. Therefore, in the case of this species the fertilization procedure should involve delayed sperm addition – 15–30 s following eggs activation.

Activation of the Gametes

The eggs of freshwater teleosts lose their fertilizing capacity after a certain period of time following contact with water (Coward et al. 2002; Minin and Ozerova 2008). The period during which the eggs remains fertilizable is different for different species. For example, using hatchery water as the AS, the eggs of rainbow trout, *Ocnorhynchus mykiss* (Walbaum), were capable of fertilization for about 30 s (Liley et al. 2002), eggs of common carp for approximately 1 min (Żarski et al. 2015) and eggs of Eurasian perch for even 2.5 min (Żarski et al. 2012a). These differences in the duration of eggs susceptibility for fertilization creates the need for slight, but very important modifications in the fertilization procedure adjusted specifically to each species, and this also applies to perch. Interestingly, to the best of our knowledge, the duration of activity of Eurasian perch eggs is the highest among freshwater species.

Fish spermatozoa (with the exception of live-bearing species) are activated following their exposure to the aquatic environment. Three models of activation are known: spermatozoa of most freshwater fish are activated by a decrease of osmolality, those of marine species by an increase of osmolality and those of salmonids and acipenseriform fish are activated by a decrease of K^+ concentration in the extracellular space (Morisawa 2008). All these phenomena trigger a complex chain of events that ultimately lead to the engagement of a protein called dynein in the flagellum whose action is directly expressed in the motion of the flagellum. Perch spermatozoa are activated by a decrease of osmolality, thus, follow the general freshwater model of activation. The period of active movement is dependent on the concentration (and thus osmolality) of the activating solution with less than 90 s of active movement in water up to more than 120 min in a solution with the osmolality of 210 mOsmol kg^{-1} (Lahnsteiner 2011).

> **Important**
> Activation mechanisms of eggs of freshwater teleosts are still discussed and different theories were reported (for details see Coward et al. 2002). Therefore, the activation mechanism is not discussed here except for the fact that after contact with water eggs lose their fertilizing capacity. However, in case of percids, after the contact of eggs with water a cortical reaction occurs (Żarski et al. 2012b) which coincides with the activation of the egg. Therefore, it can be speculated that eggs of Eurasian perch are activated by the contact with water, although the mechanism standing behind it remains unclear.

Activating Solutions

In the hatchery practice, for many years hatchery water was used as an AS during *in vitro* fertilization. However, it was found that the duration of activity of eggs and sperm (when the eggs remain fertilizable and sperm remains motile) strictly depends on the composition of the AS (Krise et al. 1995; Żarski et al. 2012a; Cejko et al. 2013) where osmolality is the major factor determining the period of time during which gametes remain active (Boryshpolets et al. 2009; Żarski et al. 2015). Considering the fact that within the range between 0 and 300 mOsm kg^{-1} the duration of activity of both sperm and eggs is positively correlated with osmolality it seems unreasonable to use hatchery water that usually has an osmolality below 50 mOsm kg^{-1}. An additional disadvantage of using hatchery water as an AS is that its composition is different in different hatcheries and it is usually not constant even at the same site, which makes comparisons in the fertilization effectiveness between hatcheries as well as between years impossible. Therefore, it is recommended to use standardized AS-s which can significantly improve fertilization effectiveness. In case of the Eurasian perch many different AS-s were already tested with varying result (Żarski et al. 2012a; Sarosiek et al. unpublished). Compositions of some of the effectively applied AS-s are as follows (Sarosiek et al. unpublished):

1. 0.3 % urea, 0.4 % NaCl (Woynarovich solution)
2. 20 mM Tris, 40 mM NaHCO$_3$, pH 8.5
3. 10 mM HEPES, 100 mM NaCl, pH 8.0
4. 80 mM NaCl, 20 mM KCl, 10 mM Tris, pH 8.0

As it can easily be noticed, the composition of most of the AS-s tested is somehow complicated and require a certain accuracy of preparation, including buffering. This makes the application of these AS-s very difficult during regular hatchery practice. However, as considering *in vitro* fertilization effectiveness, the benefits of using standardized AS are considerably high, therefore, for the hatchery practice application of the simplest AS may be advised – the Woynarovich solution. This AS is a simple mixture of urea (karbamid) and NaCl which is very easy to prepare without the need for specific chemical compounds. Most importantly, it was proven to be very effective in controlled fertilization of Eurasian perch (Żarski et al. 2012a).

Important
1. For preparation of the Woynarovich solution, distilled water should be used. Demineralized, deionized or reverse osmosis water can also be used, instead. Using hatchery water can result in high osmolality of the AS and prevent effective activation of the gametes.
2. Osmolality above 300 mOsm kg^{-1} is too high to activate sperm and eggs properly. Application of hyperosmotic AS can therefore lower fertilization success or prevent fertilization alltogether.

Sperm-to-Egg Ratio

It has been proven for many fish species that one of the most important parameters affecting fertilization success is the sperm-to-egg ratio (Rinchard et al. 2005; Butts et al. 2009). Unfortunately, in case of hatchery practice this aspect is very often ignored and the amount of sperm is very often adjusted 'visually' (subjectively). However, both, excess and deficit of the sperm may reduce the effectiveness of the entire operation of controlled reproduction. The usage of too much sperm usually creates the necessity of more frequent sperm stripping which in turn, causes the 'over-exploitation' of the male stock with a direct effect on their condition and, consequently, can lead to an increased mortality of the fish. This can be overcome by using higher numbers of males during the spawning period. This can, however, induce much higher costs of induced spawning. On the other hand, the application of too low volumes of sperm during the fertilization process is usually the reason of lower fertilization rates (Rinchard et al. 2005; Linhart et al. 2006; Litvak et al. 2008) that has a direct negative effect on the number of larvae produced. Therefore, application of the proper amount of sperm should be considered as a standard hatchery protocol to which a great deal of attention should be paid.

It was found that the optimal sperm-to-egg ratio amounts to 250,000 of spermatozoa to 1 egg in the Eurasian perch, when sperm with motility of over 80 % is used (Bernáth et al. unpublished). Concentration of perch spermatozoa ranges between 3.3 and 66.5×10^9 per ml (Table 9.1). It is then difficult to give clear, uniform recommendation how much sperm is necessary to use in the hatchery practice. Considering the abovementioned and the fact that in the case of Eurasian perch the number of dry eggs per gram may range between 350 and 700 (depending on the female's size, Bernáth et al. unpublished, Żarski et al. unpublished), it is

Table 9.1 Sperm concentrations (range or mean) of Eurasian perch according to different authors

| Broodstock | Sperm concentration ($\times 10^9$ per ml) | | | Reference |
	Min	Max	Mean ±SD	
Domesticated[a]	27.9	42.0	–	Rougeot et al. (2004)
Domesticated	29.8	39.1	–	
Wild	21.7	32.6	–	Król et al. (2006)
Wild	13	19	–	Lahnsteiner et al. (1995)
Pond-reared	3.3	43.9	–	Alavi et al. (2007)
Pond-reared	36.0	66.5	–	Alavi et al. (2010)
Domesticated[b]	–	–	45.3 ± 5.4	Rodina et al. (2008)
Domesticated	–	–	37.8 ± 6.3	
Pond reared	29.6	37.8	–	Żarski et al. (unpublished)
Wild	23.4	45.2	–	Bernáth et al. (unpublished)

[a]Sperm obtained from hormonally sex-reversed males (neomales) was analysed
[b]Testicular sperm obtained from hormonally sex-reverrsed males (neomales) was used

recommended to verify sperm concentration as well as the number of eggs per g (in the dry ribbon) and follow the recommendations given in the Table 9.2.

> **Important**
> The values presented in Table 9.2 are the critical, minimum volumes (threshold values) of sperm which ensure the fertilization success and are more suitable for scientific purposes. For commercial purposes we highly recommend to use 1 ml of freshly collected sperm per each 100 g of dry eggs, what should allow to ensure high fertilization rate. However, in that case extremely low sperm concentration should be considered as a possible limiting factor and the reason for any fertilization failure.

Practical Aspects of *In Vitro* Fertilization

<u>Necessary items</u>:

- Collected dry eggs
- Collected sperm (from at least three males)
- Precise scales (with accuracy of ±1 g)
- Plastic bowl for fertilization
- Benchtop
- Activating solution (it is recommended to use Woynarovich solution: 0.4 g NaCl + 0.3 g urea per 1 l of distilled water) with the temperature close (±2 °C) to the one in the incubation unit
- Hatchery water (from the incubation unit)
- Incubators and incubating unit (see Chap. 10)

Table 9.2 The volume of sperm (ml of sperm per 100 g of dry eggs) recommended to use for *in vitro* fertilization of Eurasian perch eggs in relation to sperm concentration ($\times 10^9$ of spermatozoa per ml) and number of dry eggs per g

Number of eggs per gram	Sperm concentration ($\times 10^9$ of spermatozoa per ml)												
	5	10	15	20	25	30	35	40	45	50	55	60	65
350	1.75	0.88	0.58	0.44	0.35	0.29	0.25	0.22	0.19	0.18	0.16	0.15	0.13
400	2.00	1.00	0.67	0.50	0.40	0.33	0.29	0.25	0.22	0.20	0.18	0.17	0.15
450	2.25	1.13	0.75	0.56	0.45	0.38	0.32	0.28	0.25	0.23	0.20	0.19	0.17
500	2.50	1.25	0.83	0.63	0.50	0.42	0.36	0.31	0.28	0.25	0.23	0.21	0.19
550	2.75	1.38	0.92	0.69	0.55	0.46	0.39	0.34	0.31	0.28	0.25	0.23	0.21
600	3.00	1.50	1.00	0.75	0.60	0.50	0.43	0.38	0.33	0.30	0.27	0.25	0.23
650	3.25	1.63	1.08	0.81	0.65	0.54	0.46	0.41	0.36	0.33	0.30	0.27	0.25
700	3.50	1.75	1.17	0.88	0.70	0.58	0.50	0.44	0.39	0.35	0.32	0.29	0.27

Fertilization procedure:

1. Place the eggs from the holding container(s) (in which the eggs were stored) into a larger dry bowl. The size of the bowl depends on the number of egg-ribbons about to be fertilized, nonetheless the size of the bowl should allow to freely mix the eggs inside by hand.
2. Pour the activating solution (AS) onto the eggs in the way that the eggs can be freely dispersed within the entire volume of the AS. Mix the eggs gently for 15 s to let them disperse in the bowl.
3. After 15–30 s following egg activation (the bigger the egg ribbon or the more ribbons are fertilized – the later should sperm be added) add the sperm to the bowl and continue to gently mix the eggs for the next 2 min
4. Pour the AS from the bowl (do not remove the eggs) and fill the bowl with freshwater (from the incubation unit)
5. Leave the eggs for 30 min in the bowl and pay attention to keep a constant temperature (the temperature should not increase for more than 2 °C) by exchanging water in the bowl, if necessary
6. 30 min later (which allows the eggs to water-harden and acquire handling resistance) transfer the eggs gently into the incubators

Practical Advice

It is adviced to have always approx. 2 ml of stripped sperm (either with or without extender) already prepared before the females are checked for ovulation. This way when the fish will release eggs spontaneously during manipulations, the eggs can immediately be moved into the bowl and fertilized with the prepared sperm. It is important, that sperm can be stored for a relatively long period of time (with or without the application of extenders – see Chap. 7) and eggs remain active for about 2.5 min after contact with hatchery water. This creates a real possibility to effectively fertilize eggs released during manipulation.

References

Alavi SMH, Rodina M, Policar T, Kozak P, Psenicka M, Linhart O (2007) Semen of Perca fluviatilis L.: sperm volume and density, seminal plasma indices and effects of dilution ratio, ions and osmolality on sperm motility. Theriogenology 68:276–283. doi:10.1016/j.theriogenology.2007.05.045

Alavi SMH, Rodina M, Hatef A, Stejskal V, Policar T, Hamáčková J, Linhart O (2010) Sperm motility and monthly variations of semen characteristics in Perca fluviatilis (Teleostei: Percidae). Czech J Anim Sci 55:174–182

Billard R, Petit J, Jalabert B, Szollosi D (1974) Artifical insemination in trout using a sperm diluant. In: Blaxter JHS (ed) The early life history of fish. Springer, Berlin/Heidelberg, pp 715–723

Boryshpolets S, Dzyuba B, Stejskal V, Linhart O (2009) Dynamics of ATP and movement in Eurasian perch (Perca fluviatilis L.) sperm in conditions of decreasing osmolality. Theriogenology 72:851–859. doi:10.1016/j.theriogenology.2009.06.005

Butts IAE, Trippel EA, Litvak MK (2009) The effect of sperm to egg ratio and gamete contact time on fertilization success in Atlantic cod Gadus morhua L. Aquaculture 286:89–94. doi:10.1016/j.aquaculture.2008.09.005

Cejko BI, Sarosiek B, Kowalski RK, Krejszeff S, Kucharczyk D (2013) Application of computer-assisted sperm analysis in selecting the suitable solution for common carp, Cyprinus carpio L., sperm motility. J World Aquacult Soc 44:466–472. doi:10.1111/jwas.12043

Coward K, Bromage NR, Hibbitt O, Parrington J (2002) Gamete physiology, fertilization and egg activation in teleost fish. Rev Fish Biol Fish 12:33–58. doi:10.1023/A:1022613404123

Krise WF, Hendrix MA, Bonney WA, Baker-Gordon SE (1995) Evaluation of sperm-activating solutions in Atlantic salmon Salmo salar fertilization tests. J World Aquacult Soc 26:384–389. doi:10.1111/j.1749-7345.1995.tb00833.x

Król J, Glogowski J, Demska-Zakes K, Hliwa P (2006) Quality of semen and histological analysis of testes in Eurasian perch Perca fluviatilis L. during a spawning period. Czech J Anim Sci 51:220–226

Lahnsteiner F (2011) Spermatozoa of the teleost fish Perca fluviatilis (perch) have the ability to swim for more than two hours in saline solutions. Aquaculture 314:221–224. doi:10.1016/j.aquaculture.2011.02.024

Lahnsteiner F, Berger B, Weismann T, Patzner RA (1995) Fine structure and motility of spermatozoa and composition of the seminal plasma in the perch. J Fish Biol 47:492–508. doi:http://dx.doi.org/10.1006/jfbi.1995.0154

Liley NR, Tamkee P, Tsai R, Hoysak DJ (2002) Fertilization dynamics in rainbow trout (Oncorhynchus mykiss): effect of male age, social experience, and sperm concentration and motility on in vitro fertilization. Can J Fish Aquat Sci 152:144–152. doi:10.1139/f01-202

Linhart O, Rodina M, Kocour M, Gela D (2006) Insemination, fertilization and gamete management in tench, Tinca tinca (L.). Aquac Int 14:61–73. doi:10.1007/s10499-005-9014-1

Litvak MK, Butts IAE, Rideout RM (2008) Cryopreservation of sperm from Winterflounder, Pseudopleuronectes americanus. In: Methods in reproductive aquaculture. CRC Press, Boca Raton, pp 459–462

Minin AA, Ozerova SG (2008) Spontaneous activation of fish eggs is abolished by protease inhibitors. Russ J Dev Biol 39:293–296. doi:10.1134/S1062360408050056

Morisawa M (2008) Adaptation and strategy for fertilization in the sperm of teleost fish. J Appl Ichthyol 24:362–370. doi:10.1111/j.1439-0426.2008.01126.x

Radziwoński J (1852) O sztucznem zapładnianiu ikry rybiej w zastosowaniu do chowu pstrągów: rzecz czytana na posiedzeniu Towarzystwa nauk

Rinchard J, Dabrowski K, Van Tassell JJ, Stein RA (2005) Optimization of fertilization success in Sander vitreus is influenced by the sperm: egg ratio and ova storage. J Fish Biol 67:1157–1161. doi:10.1111/j.0022-1112.2005.00800.x

Rodina M, Policar T, Linhart O, Rougeot C (2008) Sperm motility and fertilizing ability of frozen spermatozoa of males (XY) and neomales (XX) of perch (Perca fluviatilis). J Appl Ichthyol 24:438–442. doi:10.1111/j.1439-0426.2008.01137.x

Rougeot C, Nicayenzi F, Mandiki SNM, Rurangwa E, Kestemont P, Mélard C (2004) Comparative study of the reproductive characteristics of XY male and hormonally sex-reversed XX male Eurasian perch, Perca fluviatilis. Theriogenology 62:790–800. doi:10.1016/j.theriogenology.2003.12.002

Woynarovich E, Woynarovich A (1980) Modified technology for elimination of stickiness of common carp Cyprinus carpio eggs. Aquac Hung 2:19–21

Żarski D, Horváth Á, Kotrik L, Targońska K, Palińska K, Krejszeff S, Bokor Z, Urbányi B, Kucharczyk D (2012a) Effect of different activating solutions on the fertilization ability of Eurasian perch, Perca fluviatilis L., eggs. J Appl Ichthyol 28:967–972. doi:10.1111/jai.12098

Żarski D, Krejszeff S, Palińska K, Targońska K, Kupren K, Fontaine P, Kestemont P, Kucharczyk D (2012b) Cortical reaction as an egg quality indicator in artificial reproduction of pikeperch, Sander lucioperca. Reprod Fertil Dev 24:843. doi:10.1071/RD11264

Żarski D, Horváth Á, Bernáth G, Palińska-Zarska K, Krejszeff S, Müller T, Kucharczyk D (2014) Application of different activating solutions to in vitro fertilization of crucian carp, Carassius carassius (L.), eggs. Aquac Int 22:173–184. doi:10.1007/s10499-013-9692-z

Żarski D, Cejko BI, Krejszeff S, Palińska-Żarska K, Horvath A, Sarosiek B, Judycka S, Kowalski RK, Łączyńska B, Kucharczyk D (2015) The effect of osmolality on egg fertilization in common carp, Cyprinus carpio Linnaeus, 1758. J Appl Ichthyol 31:159–163. doi:10.1111/jai.12739

Żarski D, Bernath G, Król J, Cejko BI, Bokor Z, Palińska-Żarska K, Milla S, Fontaine P, Krejszeff S. Hormonal manipulation improves spermiation in Eurasian perch, Perca fluviatilis L., a freshwater teleostei fish. unpublished

Theoretical Background

Incubation of the developing embryos and control over the hatching process is the final step of controlled reproduction and the first step towards the control of fish life cycle (Schaerlinger and Żarski 2015). Incubation involves all methods that provide optimal conditions for the embryos to develop until they reach the hatching stage. In other freshwater species, such as for example pikeperch, the incubation process must be preceded by the specific procedures of removal of eggs adhesiveness, which is usually crucial for successful incubation (Zakęś and Demska-Zakęś 2009; Żarski et al. 2015b). However, in the case of Eurasian perch, due to specific characteristics of its eggs (spawned in the form of a ribbon), there is no need for any treatment following fertilization. Generally, the main concern during the incubation of Eurasian perch egg-ribbons is to provide suitable circulation of well oxygenated water around all the eggs. For that purpose many kinds of different incubators may be used (Kucharczyk et al. 1996; Żarski et al. 2011).

Hatching, is the set of procedures leading to acquiring hatched larvae without any egg debris and ready to transfer to the rearing unit. During hatching, special attention should be paid to removing the structure of the egg ribbon (jelly-like structures) in the most effective way. This is actually a crucial aspect since the ribbon structure loosens just before the larvae are ready to hatch (Formicki et al. 2009) and becomes a 'jelly-like' shapeless mass that is difficult to remove. Therefore, very careful supervision over the hatching process is highly advised. Especially, that hatching in percids is usually an asynchronous event lasting between 3 days at 19 °C and 5 days at 15 °C even if the same egg batch is incubated (Żarski et al. 2011, 2015a).

© The Authors 2017
D. Żarski et al., *Controlled Reproduction of Wild Eurasian Perch*, SpringerBriefs in Environmental Science, DOI 10.1007/978-3-319-49376-3_10

Embryonic Development

At 12–13 °C, the first cell division can be observed within 3 h after fertilization. At 24 h following fertilization, the embryos should reach the blastula stage which is followed by the onset of the epiboly process heralding the beginning of gastrulation. This should be completed after the first 2 days of incubation, when the blastoderm completely covers the yolk (100 % epiboly). The subsequent steps are the organo-genesis and tail segmentation that last until pigmentation of the eyes. The specific moment when pigmentation is clearly visible in the eyes of embryos takes place just before hatching in the Eurasian perch. At this moment, morphogenesis of many organ rudiments is almost complete and slows down considerably (Fig. 10.1).

Verification of the Effectiveness of Fertilization

Within the first 3 days of incubation, a significant embryonic mortality may be observed (Alix et al. 2013). Therefore, the success of fertilization should not be determined earlier than 72 h following fertilization, when the embryos reach the late neurula stage (when the body of the embryo can already be observed on the animal pole of the embryo – as described by Iwamatsu 2004). However, it should be high-lighted, that at this stage the most common developmental abnormalities of embryo cannot be recognized, yet. Considering the fact that the hatching rate is usually much lower than the fertilization rate determined at earlier stages (Żarski et al. 2011; Alix et al. 2013; Schaerlinger and Żarski 2015), fertilization success (being

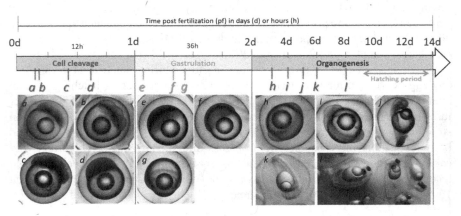

Fig. 10.1 Development of the Eurasian Perch (P. fluviatilis). The general features characterized by observation performed in living embryos during incubation at 13 °C (For more details see Alix et al. 2015). Description of the developmental steps: *a – 2 cells (3.5 hpf); b – 4 cells (4 hpf); c – 128 cells (10 hpf); d – High blastula (15 hpf); e – 30 % epiboly (26 hpf); f – 50 % epiboly (31 hpf); g – 90 % epiboly (41 hpf); h – Optic capsule (66 hpf); i – Otic vesicle (96 hpf); j – Tail elongation (5 dpf); k – Eye pigmentation onset (6 dpf); l – Eyed-egg stage (8 dpf)* (Photo: *a, k, l* – Palińska-Żarska K.; *b–j* – by courtesy of Alix M)

the overall result of gamete quality stemming from controlled reproduction procedures as well as fertilization effectiveness) should be verified at hatching. However, it should be emphasized that even those larvae that are able to hatch can be characterized by varying quality, as highly deformed individuals were also reported to hatch spontaneously (Żarski et al. 2011; Alix et al. 2013; Schaerlinger and Żarski 2015). Castets et al. (2012) indicated that the resistance of the larvae to starvation as well as their survival at day 7 post hatching were the indicators characterized highest egg quality. This additionally confirms that the larval performance can be more reliable egg quality indicator than fertilization and/or hatching rate. Therefore, a recently reported study suggests that the most reliable method for the verification of reproductive success in Eurasian perch is the determination of number of larvae with inflated swim bladder (Żarski et al. 2015c). This method of verification of the spawning outcome excludes from the overall mass of the larvae produced those with low biological quality whith the highest accuracy. This stems from the fact that the larvae characterized by lower quality are in most cases unable to inflate their swim bladders, which is a crucial event in the fish lifecycle (Woolley and Qin 2010; Palińska-Żarska et al. 2014). Therefore, to verify the productivity of spawning (in both scientific research and commercial production) the use of spawning efficiency index (SEI) is recommended to be used. This index is calculated by the number of larvae with inflated swim bladder in relation to body weight of the females spawned and returns the global overview on the final quality of the larvae stemming from the effectiveness of the entire controlled reproduction operation.

Spawning Efficiency Index (SEI)
The Spawning Efficiency Index developed for Eurasian perch (Żarski et al. 2015c) involves the determination of the number of larvae with inflated swim bladder in relation to the 1 kg of body weight of the females spawned.

$$SEI = N / BWF$$

where:

N – the number of larvae with inflated swim bladder
BWF – the body weight of the spawned females (in kg).

Practical Advice
Non-developing eggs should be removed as soon as low egg quality is recognized. If it is impossible to recognize earlier, dead ribbons (or dead parts of the ribbons) should be removed if only white (non-transparent) eggs are present upon macroscopic observation. This will allow to keep the rearing unit in good sanitary conditions.

Incubation Devices

In fact, there are no standardized incubators for Eurasian perch eggs. For the purpose of incubation many different devices were constructed and tested and most of them were found to be effective. From the commercial point of view, however, the most suitable are those incubators that allow to 'suspend' the egg ribbons in the water column, which is very often the rearing unit intended to be used for initial rearing of larvae (Fig. 10.2). For this purpose, different floating cages or specific trays (floating or fixed just above the water surface) with the bottom replaced with a net (or any other riddled bottom) can be recommended (Fig. 10.3). In these incubators eggs can be washed from all sides allowing proper water exchange around all the eggs in the ribbon. However, it is important to provide the mesh size (or size of the holes in the bottom) that is large enough (about 3.0 mm) to allow freshly hatched

Fig. 10.2 Egg ribbons obtained from different females incubated in a separate floating 'cages' in a small scale experimental RAS, in which larvae were reared after hatching (Photo: D. Żarski)

Fig. 10.3 An example of an incubator for Eurasian perch egg ribbons with bottom, and part of the one wall replaced with a net (Photo: D. Żarski)

larvae to fall to the bottom of the tank, when transferred from the incubation to the rearing unit. However, the mesh should not be too large either, as ribbon debris can also fall through. The mesh size should allow to separate the larvae easily from the ribbon leftovers just by removing the cages (trays) from the tank (incubation unit).

Practical Advice
In order to facilitate the work with eggs it is recommended to use one hatchery unit for the incubation of all eggs, and transfer the eggs to the rearing system just prior to hatching. In this case, the temperature can be about 2–3 °C higher in the rearing unit. This will allow to synchronize the hatching process of the batches designated to a particular rearing unit and to minimize the amount of possible debris and/or pathogens originating from the low quality eggs in the rearing units.

Incubation Conditions

During the incubation process, apart from the oxygen dissolved in the water, another very important parameter to provide is adequate temperature and pH. In addition to the fact that temperature directly influences the developmental rate of the embryos, too low or too high temperature may also induce improper embryonic development leading to developmental abnormalities and/or embryo mortality (Kamler 2002). It was found that too low pH (below 5) can also be the cause for a prolonged

Table 10.1 Water quality requirements to be used in the hatchery unit

Dissolved oxygen (DO)	>3.5 ppm
CO_2	<5.0 ppm
pH	6.5–9.0
Calcium	10–160 ppm
Phosphorus	0.01–3.00 mg L^{-1}
Total hardness	50–400 ppm
Hydrogen sulfide	0
Nitrite (NO_2)	<0.1 mg L^{-1}
Unionized ammonia (NH_3)	< 0.0125 mg L^{-1}

According to Hart et al. (2006)

incubation period and embryo mortality (Rask 1983). Generally in case of the Eurasian perch, eggs can successfully be incubated at a thermal range between 8 and 18 °C, with 12.5 °C suggested to be the optimal one (Teletchea et al. 2009; Żarski et al. 2015a) at which the incubation lasts approx. 165 degree-days (about 13 days). However, the most commonly applied temperature for egg incubation of Eurasian perch ranges between 12 and 16 °C (e.g. Żarski et al. 2015a), which can be recommended to apply in the hatchery practice.

Generally, apart from the parameters such as temperature, pH and dissolved oxygen in the water there are no specific requirements for the incubation of Eurasian perch eggs. Therefore, water parameters in the incubation unit should meet the typical criteria for freshwater fishes. In the most ideal situation, the water should meet the general criteria for 'drinking water' authorized to be used for human consumption. However, if it is not possible to provide such a high-quality water, a special attention should be paid to using water that meets the typical criteria for fish hatcheries. In this regards the requirement given by Hart et al. (2006) for yellow perch, Perca flavescens, a closely related fish species to Eurasian perch, can be followed (Table 10.1). Overall, the water source should be free of any suspensions (transparent), odors as well as any dissolved toxic substances (including nitrogen and phosphorus compounds very often accumulating in RAS systems). In case of using open waters for supplying the hatchery unit it is important to clarify the water (to remove all the suspended particles such as mud, organic matter etc.) by mechanical filtration (e.g. with the use of drum filters) and effectively sterilize it prior to using for supplying the system. Special attention should also be paid to the direct surroundings of the water source (lake or river) to see whether intensive agricultural, industrial or even domestic pollutants will not penetrate the water with harmful substances. In case of using well water, special attention should be paid to purification of the water. Depending on the quality, it may be necessary to remove the iron and/or manganese compounds, decalcify and/or remove other excessively contained compounds prior to use. If tap water has to be used, it is important to make sure that it was properly dechlorinated before any use in the entire hatchery unit.

Important

Eggs should not lay on the bottom of the tank during incubation and should be maximally extended. This is due to the fact that from the eyed-egg stage the embryos may start to die due to their increased oxygen demand before hatching and looseness of the jelly-like structure making oxygen exchange more difficult.

Hatching

Prior to hatching, larvae develop hatching gland cells which are responsible for the production of a hatching enzyme called 'chorionase' (Luczynski et al. 1987; Rechulicz 2001). This enzyme allows to gently digest the internal structure of the chorion surrounding the developing embryo and as a results, the corion loses its rigidity. This, together with the loosening phenomenon of the 'jelly-like' structure of the external layer of the embryo (*zona radiata externa*) just prior to hatching (Formicki et al. 2009) allows the larvae to exit the eggs easily. This corresponds with a very active movement of the larvae inside the egg which provides additional help for the larvae to leave the egg envelope. It is very important to note, that the number and size of the hatching gland cells is also dependent on the water temperature where the highest chorionase production capacity is speculated to occur usually at optimal incubation temperatures (Rechulicz 2001). Although this requires further investigations in the Eurasian perch, it additionally justifies that the 'optimal thermal range' should be maintained during the incubation period.

Considering the fact, that the hatching process is dependent on the enzymatic activity, it is therefore recommended to assist hatching by increasing the water temperature which, in turn, will increase the enzymatic activity of chorionase. In the hatchery practice this involves an increase of water temperature by a few degrees (maximum up to 5 °C) when the first spontaneously hatched larvae are observed. This can shorten the hatching process and allow most of the larvae to hatch at the time of the mouth opening or shortly before (Alix et al. 2015) (Fig. 10.4).

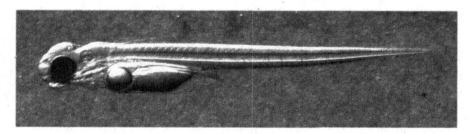

Fig. 10.4 Eurasian perch just after hatching (5.71 mm total length, incubated at 14 °C) (Photo: K. Palińska-Żarska)

Practical Advice
1. It is recommended to incubate the eggs at 12–13 °C and to support the hatching process with an increase of the temperature to maximum 15 °C.
2. Do not apply thermal treatment to the eggs when parallel incubation of the eggs obtained at different dates is conducted in the same hatchery unit.

Initial Larvae Rearing (General Advices)

After hatching larvae are recommended to be kept at 15 °C for first 14 days (Palińska-Żarska et al. unpublished). During that time the light conditions should be adjusted in such a way, that larvae are scattered all over the tank or they are gathering in the middle of the tank just under the water surface. It should be avoided to provide light conditions inducing gathering the larvae close to the tank walls and/or in the corners of the tank, what can alter the swim bladder inflation process (Palińska-Żarska et al. 2013). At such a temperature the first food should be offered between 5 and 7 days following hatching. As the first source of food larvae should have offered freshly hatched *Artemia* sp. nauplii.

References

Alix M, Schaerlinger B, Ledoré Y, Chardard D, Fontaine P (2013) Developmental staging and deformities characterization of the Eurasian perch, Perca fluviatilis. Commun Agric Appl Biol Sci 78:12–14

Alix M, Chardard D, Ledoré Y, Fontaine P, Schaerlinger B (2015) An alternative developmental table to describe non-model fish species embryogenesis: application to the description of the Eurasian perch (Perca fluviatilis L. 1758) development. Evodevo 6:39. doi:10.1186/s13227-015-0033-3

Castets M-DD, Schaerlinger B, Silvestre F, Gardeur J-NN, Dieu M, Corbier C, Kestemont P, Fontaine P (2012) Combined analysis of Perca fluviatilis reproductive performance and oocyte proteomic profile. Theriogenology 78(432–42):442–13. doi:10.1016/j.theriogenology.2012.02.023

Formicki K, Smaruj I, Szulc J, Winnicki A (2009) Microtubular network of the gelatinous egg envelope within the egg ribbon of European Perch, Perca Fluviatilis L. Acta Ichthyol Piscat 39:147–151. doi:10.3750/AIP2009.39.2.10

Hart SD, Garling DL, Malison JA (2006) Yellow perch (Perca flavescens) culture guide. North Central Regional Aquaculture Center, Iowa State University, Ames

Iwamatsu T (2004) Stages of normal development in the medaka Oryzias latipes. Mech Dev 121:605–618. doi:10.1016/j.mod.2004.03.012

Kamler E (2002) Ontogeny of yolk-feeding fish: an ecological perspective. Rev Fish Biol Fish 12:79–103. doi:10.1023/A:1022603204337

Kucharczyk D, Kujawa R, Mamcarz A (1996) New experimental incubation unit for eggs of the Perch Perca fluviatilis. Progress Fish Cult 58:281–283. doi:10.1577/1548-8640(1996)058<0281:NEIUFE>2.3.CO;2

Luczynski M, Strzczek J, Brzuzan P, Łuczyński M, Strzeżek J, Brzuzan P (1987) Secretion of hatching enzyme and its proteolytic activity in coregoninae (Coregonus albula L. and Coregonus-Lavaretus L.) embryos. Fish Physiol Biochem 4:57–62. doi:10.1007/bf02044314

Palińska-Żarska K, Krejszeff S, Łopata M, Żarski D. Effect of temperature and tank wall colour on the effectiveness 1 of swim bladder inflation in Eurasian Perch, Perca fluviatilis L., larvae reared under controlled conditions (unpublished)

Palińska-Żarska K, Żarski D, Krejszeff S, Nowosad J, Biłas M, Kucharczyk D (2013) Tank wall color affects swimbladder inflation in Eurasian perch, Perca fluviatilis l., under controlled conditions. Commun Agric Appl Biol Sci 78:338–341

Palińska-Żarska K, Żarski D, Krejszeff S, Nowosad J, Biłas M, Trejchel K, Kucharczyk D (2014) Dynamics of yolk sac and oil droplet utilization and behavioural aspects of swim bladder inflation in burbot, Lota lota L., larvae during the first days of life, under laboratory conditions. Aquac Int 22:13–27. doi:10.1007/s10499-013-9663-4

Rask M (1983) The effect of low pH on perch, Perca fluviatilis L. I. Effects of low pH on the development of eggs of perch. Ann Zool Fenn 20:73–76

Rechulicz J (2001) Incubation temperature effects on the development of hatching gland cells in ide, Leuciscus idus (L.). Electron J Pol Agric Univ 4:#03

Schaerlinger B, Żarski D (2015) Evaluation and improvements of egg and larval quality in percid fishes. In: Kestemont P, Dabrowski K, Summerfelt RC (eds) Biology and culture of percid fishes. Springer, Dordrecht, pp 193–223

Teletchea F, Fostier A, Kamler E, Gardeur J-N, Le Bail P-Y, Jalabert B, Fontaine P (2009) Comparative analysis of reproductive traits in 65 freshwater fish species: application to the domestication of new fish species. Rev Fish Biol Fish 19:403–430. doi:10.1007/s11160-008-9102-1

Woolley LD, Qin JG (2010) Swimbladder inflation and its implication to the culture of marine finfish larvae. Rev Aquac 2:181–190. doi:10.1111/j.1753-5131.2010.01035.x

Zakęś Z, Demska-Zakęś K (2009) Controlled reproduction of pikeperch Sander lucioperca (L.): a review. Arch Pol Fish 17:153–170. doi:10.2478/v10086-009-0014-z

Żarski D, Palińska K, Targońska K, Bokor Z, Kotrik L, Krejszeff S, Kupren K, Horváth Á, Urbányi B, Kucharczyk D (2011) Oocyte quality indicators in Eurasian perch, Perca fluviatilis L., during reproduction under controlled conditions. Aquaculture 313:84–91. doi:10.1016/j.aquaculture.2011.01.032

Żarski D, Horváth A, Held JA, Kucharczyk D (2015a) Artificial reproduction of percid fishes. In: Kestemont P, Dąbrowski K, Summerfelt RC (eds) Biology and culture of percid fishes, 1st edn. Springer, Dordrecht, pp 123–161

Żarski D, Krejszeff S, Kucharczyk D, Palińska-Żarska K, Targońska K, Kupren K, Fontaine P, Kestemont P (2015b) The application of tannic acid to the elimination of egg stickiness at varied moments of the egg swelling process in pikeperch, Sander lucioperca (L.). Aquac Res 46:324–334. doi:10.1111/are.12183

Żarski D, Krejszeff S, Palińska-Żarska K, Bernath G, Urbanyi B, Bokor Z (2015c) The spawning efficiency index as a tool in aquaculture research and production. In: Poleksic V, Markovic Z (eds) 7th international conference "WATER & FISH." University of Belgrade, Faculty of Agriculture, Belgrade, pp 23–28

Advanced Spawning (Out-of the Season Spawning)

Theoretical Background

The economic feasibility of intensive commercial aquaculture production of Eurasian perch is dependent on the ability of supplying the market all year round with a high quality product, while also meeting the criteria of a standardized size. However, Eurasian perch spawns only once a year with the spring as its natural spawning season (Treasurer 1981; Treasurer and Holliday 1981; Długosz 1986; Sulistyo et al. 1998). Therefore, in order to sustain a continuous production throughout the year it is necessary to spawn the fish out-of the spawning season (Migaud et al. 2002).

In case of a domesticated broodstock kept in a fully controlled environment (including photo-thermal regime, water quality parameters, feeding etc.) spawning can theoretically be induced at any time of the year, if adequate stimulation of the gonadal development and spawning is provided (Fontaine et al. 2006; Abdulfatah et al. 2011, 2013; Fontaine et al. 2015). Although in this case reproduction is literally performed out-of natural spawning season of Eurasian perch, the fish are subjected to a 'steered' environmental photo-thermal cycle intended to simulate the natural annual cycle of day lengths and temperatures. In this case, domesticated fish are usually spawned following their 'natural' photo-thermal requirements which correspond to the 'spawning season' for this particular group of fish. However, in case of the first generation kept in captivity (the fish obtained from wild progenitors during the spawning season), the spawning procedure is usually preceded by a shortening or elongation of the first gonadal development process in order to spawn the fish beyond the period of the year when they naturally hatched. Nonetheless, even in this situation fish are usually reared since the very beginning in a completely controlled environment which, in our opinion, makes the term 'spawning season' imprecise to apply. For the following spawning acts of these fish, the photo-thermal regimes are usually controlled so that they mimic the natural annual cycle, making the usage of the term 'out-of season spawning' relatively moot. Especially, that in the literature, the term 'out-of season spawning' also refer to the wild fish spawned

© The Authors 2017
D. Żarski et al., *Controlled Reproduction of Wild Eurasian Perch*, SpringerBriefs in Environmental Science, DOI 10.1007/978-3-319-49376-3_11

up to several months before the spawning season, which is actually 'out-of the spawning season'. However, the reproduction of the cultured fish is possible throughout the year, regardless the real date of the spawning season, and it allows actually to obtain the larvae at any time out-of the spawning season and not just few months before. Therefore, in order to avoid any terminological ambiguities in the present manual we have decided to specify these terms, where 'out-of season spawning' always refer to the reproduction of domesticated fish (RAS-reared), whereas for the wild fish we have introduced the term 'advanced spawning'. Of course, theoretically, the reproduction of wild fish could also be moved out-off the spawning season by extending the 'wintering period' and perform 'delayed spawning', however in this manual this aspect was omitted due to the lack of sufficient data on the possible protocol and efficiency of such a practice. Additionally, considering the aim of the present manual the aspects of 'out-of season spawning' of RAS-reared fish were omitted in this book. To those interested, essential information on the protocols and factors allowing successful reproduction of domesticated stocks can be found in Fontaine et al. (2015).

Farmers about to perform 'advanced spawning' must take into consideration the necessity of the proper completion of the process of gonadal development between two spawning acts, which takes about 1 year in the wild and involves photo-thermal fluctuations which regulate the hormonal cycle (Sulistyo et al. 1998). This, in turn, regulates the adequate course of particular phases of gonadal development which is an indispensable condition allowing production of high quality gametes (revised extensively by Fontaine et al. 2015). In addition to the fact that in natural conditions this process lasts for about 1 year, it was found that it is possible to shorten this period for a few months with the application of strictly controlled photo-thermal regimes, very often followed by an application of hormonal treatment, which is in most cases a necessary element of successful 'advanced spawning' (Szczerbowski et al. 2009; Targońska et al. 2014).

State-of-the-Art of 'Advanced Spawning' in the Eurasian Perch

In wild and pond-reared Eurasian perch, successful 'advanced spawning' is possible 4 months before the spawning season, earliest (Szczerbowski et al. 2009; Targońska et al. 2014). This is usually preceded by the capture of the fish during the Autumn (middle of October until the end of November), transferring of fish to the hatchery and application of 2 or 3 months of chilling (so called wintering period) when the fish are kept at a temperature below 10 °C (Migaud et al. 2004; Szczerbowski et al. 2009). After wintering, water temperature is usually increased to 12–14 °C and then hormonal stimulation is applied (Szczerbowski et al. 2009; Żarski et al. 2011; Targońska et al. 2014). During the chilling period fish are usually kept in a shortened photoperiod (4 L:20D, Szczerbowski et al. 2009) or constant darkness (Żarski et al. 2011). The increase of the temperature following wintering was usually performed at rate below 1 °C per day (Żarski et al. 2011).

The 'advanced spawning' protocol was described by Żarski et al. (2012) and Żarski (2012), when fish were caught by drag netting under ice in the beginning of the year (2–3 months before the spawning season). The fish were caught already after a natural chilling period. After transferring to the hatchery, only thermal stimulation from 4 to 10–12 °C was necessary. This method 'advanced spawning' has the advantage that the fish stay in natural conditions until the last moment without the stress factors caused by controlled wintering in the hatchery. The spawning effectiveness of fish obtained this way is usually high, probably due to limited stress, also caused by handling and transportation, which in this case is performed at very low temperatures. However, from the commercial point of view, production based on fish obtained during the winter period has substantial limitations since successful catching is dependent on weather conditions and the overall condition of the harvested ecosystem. Additionally, according to the observations, the importance of commercial inland fisheries has been decreasing in the last decades which limits the possibilities of harvesting fish as described above to a few exceptional case.

Considering hormonal stimulation, the types and doses of spawning agents used for the induction of final oocyte maturation and ovulation during the 'advanced spawning' are usually similar to those applied during in-season spawning [for details see Żarski et al. (2015) and Chap. 5]. The hormone is also usually applied in a single injection. The only difference is the latency time which is usually longer (at least 6 days at 12 °C – due to the early stage of maturation upon injection) and more synchronized. The latter stems from the fact that during out-of season spawning the fish are usually at a more or less similar stage of maturation (before the FOM) and the divergence in maturational stages presumably occurs later (when the GV migration starts – see Chap. 4).

Spawners of Eurasian perch were usually starved while kept in the hatchery (during the chilling period and controlled reproduction operation). This is due to the fact that wild or pond-reared fish accept only live (small fodder fish) or certain types of frozen food (such as frozen bloodworms). The adaptation of the fish to commercial pelleted food is a lengthy process. Apart from the fact that keeping the fish without feeding is cost effective and helps to maintain proper sanitary conditions in the tanks and the entire system (there is no production of the feces, no food leftovers) there is no information whether feeding the fish during the winter would improve reproduction effectiveness later on. However, from the ethical reason it should be considered to offer to the fish frozen food or some fodder fish. In such a case it has to be taken into account that biological filtration at such low temperatures is highly reduced (Wortman and Wheaton 1991; Sudarno et al. 2011). Therefore, high system capacity, water exchange and/or high biofilter volume must be provided.

Important
During the chilling (wintering) period, water quality parameters must continuously be monitored, especially ammonia levels, since even starved fish excrete high volumes of ammonia.

Practical Recommendations

The practical information provided below is based on the latest results obtained during the PERCAHATCH project, which gathers several experiments into a single protocol. These experiments allowed to determine the basic conditions for successful out-of season spawning, and included the verification of suitability of the produced larvae for further rearing, up to weaned juveniles. Although the protocol was already critically evaluated (with the use of pond-reared fish as models) and it can be recommended to follow in the hatchery practice of out-of season spawning of wild Eurasian perch, it is additionally recommended to verify it first in local hatchery conditions and with the population intended to be reproduced, before any commercial production is planned.

Broodstock Management

The fish should be obtained from a lake or earthen pond when the average daily temperature falls down to 10 °C (minimum 8 °C). After capture, fish should be transferred to the hatchery unit in which the chilling (wintering) period will be applied. Before stocking the fish into the tanks, every individual should be checked for its general health status (external parasites, gills for possible parasites or fungal infections, skin incisions, condition of the fins, other body damages, etc.) and should carefully be transferred first into a separate bath with saline solution (for details see Chaps. 2 and 3) before finally being placed into a holding tank. This practice usually allows to avoid an excessive transfer of pathogens. Next, the fish should be subjected to a 7 day acclimation period, during which mortalities should carefully be monitored. All dead fish should immediately be removed from the holding tank and the possible cause of death should be determined in order to detect the pathogens responsible for the mortality and immediate treatment, if necessary. After 7 days of acclimation a representative, randomly chosen group of fish should be removed, anesthetized (as described in Chap. 3) and another health status control should be conducted. If the fish are recognized to be suitable for wintering (with a good condition) the next step – photo-thermal manipulation – should follow.

Important
During the wintering period both sexes – males and females – should be kept together, which presumably allows for mutual pheromonal stimulation.

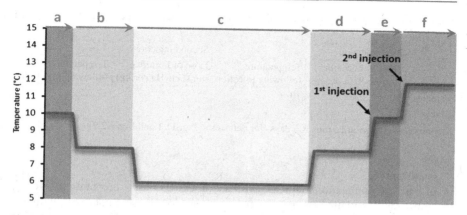

Fig. 11.1 Thermal regime (*red line*) applied before and during the wintering period (*a–e*) and spawning (*f*). The duration of this particular thermal regime: *a* – 7 days, *b* – 14 days, *c* – 40 days, *d* – 14 days, *e* – 7 days, *f* – as long as spawning occurs. *Arrows* indicate moment of hormonal injection of the females (See also Table 11.1)

Photo-Thermal Manipulation

After the acclimation period, the chilling period should be commenced. It is recommended to follow the thermal regimes presented on Fig. 11.1. Each temperature change should be carried out slowly, but not longer than 24 h (e.g. from 8 to 6 °C water temperature should be chilled within 24 h).

For the entire chilling period fish should be kept in constant dimness (without daily shifts). This allows to imitate the natural light conditions under the ice cover during wintering on a lake/pond and additionally limits stress by reducing the number of possible visual stimuli.

Stimulation of Ovulation and Spermiation

After the wintering period and the day when the hormonal treatment should be provided, another selection should be performed. For this purpose, the fish should be examined from the perspective of a health status and condition (as described above) the gender recognition should be performed. For sex recognition the protocol described in Chap. 3 should be followed.

After the fish of both sexes are recognized, the females and males should be kept separately. The hormonal stimulation of both, males and females should be conducted following the protocol given in Table 11.1.

Latency time between the second injection and ovulation should range between 32 and 108 h. However, the latency time should be verified for each population separately (for details see Chaps. 4 and 5).

Table 11.1 The protocol of hormonal stimulation of Eurasian perch during 'advanced spawning'

	First injection		Second injection	
	Dose of hormone (μg sGnRHa per kg)	Temperature following injection	Dose of hormone (μg sGnRHa per kg)	Temperature following injection
Females	10	10 °C	25	12 °C
Males	50		–	

The interval between injections is 7 days (for details see Fig. 11.1 and Fig. A2.2)

Important

In case of application of hCG (instead of GnRH-based preparations) for induction of ovulation and spermiation only one dose is recommended. In this case, fish (males and females) should be injected with 500 IU of hCG at the time of the second injection of the protocol described above (when the temperature is set to 12 °C).

References

Abdulfatah A, Fontaine P, Kestemont P, Gardeur J-N, Marie M (2011) Effects of photothermal kinetics and amplitude of photoperiod decrease on the induction of the reproduction cycle in female Eurasian perch Perca fluviatilis. Aquaculture 322–323:169–176. doi:10.1016/j. aquaculture.2011.09.002

Abdulfatah A, Fontaine P, Kestemont P, Milla S, Marie M (2013) Effects of the thermal threshold and the timing of temperature reduction on the initiation and course of oocyte development in cultured female of Eurasian perch Perca fluviatilis. Aquaculture 376–379:90–96. doi:10.1016/j. aquaculture.2012.11.010

Długosz M (1986) Oogeneza i cykl rocznego rozwoju gonad wybranych gatunków ryb w zbiornikach o odmiennych warunkach termicznych. Acta Acad Agric Techn Olst, Prot Aquarum Piscat 14:1–68

Fontaine P, Pereira C, Wang N, Marie M (2006) Influence of pre-inductive photoperiod variations on Eurasian perch Perca fluviatilis broodstock response to an inductive photothermal program. Aquaculture 255:410–416. doi:10.1016/j.aquaculture.2005.12.025

Fontaine P, Wang N, Hermelink B (2015) Broodstock management and control of reproductive cycle. In: Kestemont P, Dąbrowski K, Summerfelt RC (eds) Biology and culture of percid fishes, Principles and practices. Springer, Dordrecht, p 958

Migaud H, Fontaine P, Sulistyo I, Kestemont P, Gardeur JN (2002) Induction of out-of-season spawning in Eurasian perch Perca fluviatilis: effects of rates of cooling and cooling durations on female gametogenesis and spawning. Aquaculture 205:253–267. doi:10.1016/ S0044-8486(01)00675-5

Migaud H, Gardeur JN, Kestemont P, Fontaine P (2004) Off-season spawning of Eurasian perch Perca fluviatilis. Aquac Int 12:87–102. doi:10.1023/B:AQUI.0000017190.15074.6c

Sudarno U, Winter J, Gallert C (2011) Effect of varying salinity, temperature, ammonia and nitrous acid concentrations on nitrification of saline wastewater in fixed-bed reactors. Bioresour Technol 102:5665–5673. doi:10.1016/j.biortech.2011.02.078

Sulistyo I, Fontaine P, Rinchard J, Gardeur JN, Migaud H, Capdeville B, Kestemont P (1998) Reproductive cycle and plasma levels of steroids in male Eurasian perch Perca fluviatilis. Aquat Living Resour 11:101–110. doi:10.1016/S0990-7440(00)00146-7

Szczerbowski A, Kucharczyk D, Mamcarz A, Łuczyński MJ, Targońska K, Kujawa R (2009) Artificial off-season spawning of Eurasian perch Perca fluviatilis L. Arch Pol Fish 17:95–98

Targońska K, Szczerbowski A, Żarski D, Łuczyński MJ, Szkudlarek M, Gomułka P, Kucharczyk D (2014) Comparison of different spawning agents in artificial out-of-season spawning of Eurasian perch, Perca fluviatilis L. Aquac Res 45:765–767. doi:10.1111/are.12010

Treasurer JW (1981) Some aspects of the reproductive biology of perch Perca Fluviatilis L. Fecundity, maturation and spawning behavior. J Fish Biol 18:729–740

Treasurer JW, Holliday FGT (1981) Some aspects of the reproductive biology of perch Perca fluviatilis L. A histological description of the reproductive cycle. J Fish Biol 18:359–376. doi:10.1111/j.1095-8649.1981.tb03778.x

Wortman B, Wheaton F (1991) Temperature effects on biodrum nitrification. Aquac Eng 10:183–205. doi:10.1016/0144-8609(91)90023-D

Żarski D (2012) First evidence of pheromonal stimulation of maturation in Eurasian perch, Perca fluviatilis L., females. Turk J Fish Aquat Sci 12:771–776

Żarski D, Bokor Z, Kotrik L, Urbanyi B, Horváth A, Targońska K, Krejszeff S, Palińska K, Kucharczyk D (2011) A new classification of a preovulatory oocyte maturation stage suitable for the synchronization of ovulation in controlled reproduction of Eurasian perch Perca fluviatilis L. Reprod Biol 11:194–209. doi:10.1016/S1642-431X(12)60066-7

Żarski D, Krejszeff S, Horváth Á, Bokor Z, Palińska K, Szentes K, Łuczyńska J, Targońska K, Kupren K, Urbányi B, Kucharczyk D (2012) Dynamics of composition and morphology in oocytes of Eurasian perch, Perca fluviatilis L., during induced spawning. Aquaculture 364–365:103–110. doi:10.1016/j.aquaculture.2012.07.030

Żarski D, Horváth A, Held JA, Kucharczyk D (2015) Artificial reproduction of percid fishes. In: Kestemont P, Dąbrowski K, Summerfelt RC (eds) Biology and culture of percid fishes, 1st edn. Springer, Dordrecht, pp 123–161

Appendices

Appendix 1

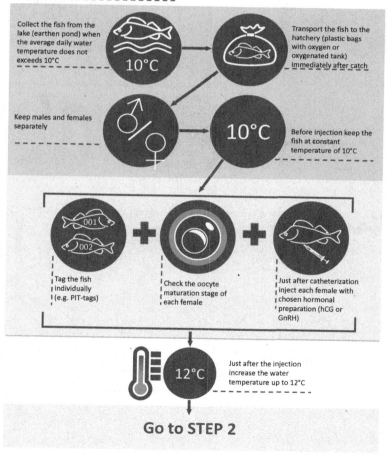

Fig. A.1.1 Schematic summary of the first step of reproductive protocol of wild Eurasian perch during the spawning season

D. Żarski et al., *Controlled Reproduction of Wild Eurasian Perch*, SpringerBriefs
in Environmental Science, DOI 10.1007/978-3-319-49376-3

STEP 2 (in-season spawning)

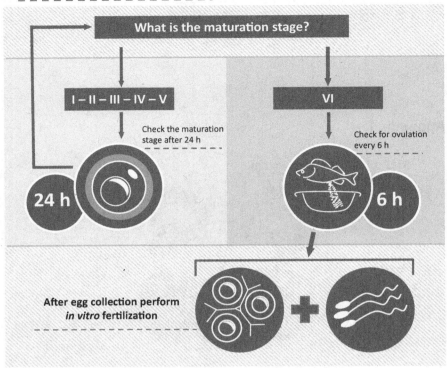

Fig. A.1.2 Schematic summary of the second (last) step of reproductive protocol of wild Eurasian perch during the spawning season

Appendix 2

STEP 1 (advanced spawning)

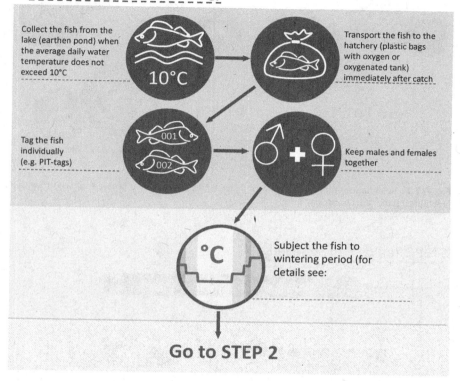

Fig. A.2.1 Schematic summary of the first step of the reproductive protocol of wild Eurasian perch during the 'advanced spawning'

STEP 2 (advanced spawning)

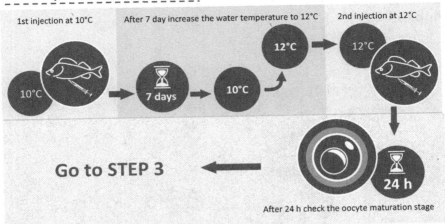

Fig. A.2.2 Schematic summary of the second step of the reproductive protocol of wild Eurasian perch during the 'advanced spawning'

STEP 3 (advanced spawning)

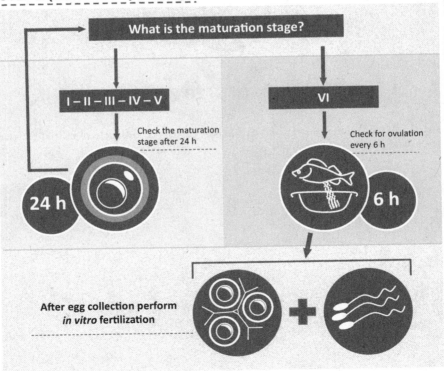

Fig. A.2.3 Schematic summary of the third (last) step of the reproductive protocol of wild Eurasian perch during the 'advanced spawning'

Printed in the United States
By Bookmasters